COSMIC CLOUDS
3-D
WHERE STARS ARE BORN

DAVID J. EICHER
BRIAN MAY Creative Director
J.-P. METSÄVAINIO 3-D Images

THE
London Stereoscopic Company,
LTD.

COSMIC CLOUDS 3-D

WHERE STARS ARE BORN

Nebula / noun — Plural noun: **nebulae**

ASTRONOMY: a cloud of gas and dust in outer space, visible in the night sky either as a bright patch or as a dark silhouette against other luminous matter.

THE MIT PRESS, CAMBRIDGE, MASSACHUSETTS

THE

London Stereoscopic Company, LTD.

**Published in 2020 by
The London Stereoscopic Company**

© The London Stereoscopic Company 2020

Text © David J. Eicher 2020

David J. Eicher has asserted his right to be identified as the author of this work in accordance with the Copyright, Design and Patents Act 1988 (UK).

The London Stereoscopic Company

Director – Brian May

Publisher and Editor-in-Chief – Robin Rees

Art Director and Illustrator – James Symonds

Astronomy Consultant and Proofreader –
  Richard Talcott

Astronomy Consultant – Glenn Smith

Archivist – Denis Pellerin

LSC Manager – Sara Bricusse

PR Manager – Nicole Ettinger

Office Manager – Sally Avery-Frost

Website – Phil Murray

Published in North America, Australia, and New Zealand by **The MIT Press**
Cambridge, MA 02142
Library of Congress Control Number: 2019948158
MIT Press ISBN 978-0-262-04402-8

A catalogue record for this book is available from the British Library.
UK ISBN 978-1-9996674-7-4
Book printed and bound in China by
Jade Productions.
OWL Stereoscope designed by Brian May.
First printing 2020.

www.LondonStereo.com
www.BrianMay.com

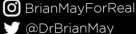 BrianMayForReal
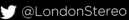 @DrBrianMay
@LondonStereo
 @LondonStereo
 @LondonStereoscopicCompany

Note about the images

Unless otherwise indicated all the 3-D photographs in this book are by J.-P. Metsävainio.
Monos are individually credited.

Every effort has been made to correctly credit the photographers who have contributed to this work. If there are any wrong attributions, please contact the publishers.

On page 2: Set within the constellation of Cygnus the Swan, the bright edge of a ring-like nebula is traced by the glow of ionized hydrogen and oxygen gas, swept by a wind from the star WR 134 (Wolf-Rayet 134), the brightest star within the ring. *Credit: J.-P. Metsävainio*

On this page: The Bug Nebula (NGC 6302) in Scorpius is a distinctive bipolar planetary nebula with a flattened shape, the gas squeezing outward from the star's poles. Located 3,400 light-years away, it is a fine challenge for observers with moderate-size telescopes.
Credit: J.-P. Metsävainio

End papers: The Milky Way over the Atacama Desert
This deep exposure brings out an array of sky wonders, including Barnard's Loop — the half pink ring visible above the Milky Way band — and the Small and Large Magellanic Clouds — the two pale diffuse clouds floating towards the upper right corner.
Credit: Petr Horálek

CONTENTS

PREFACE

Sometimes in life, if we're lucky, adventures beget adventures. The first book I collaborated on with Brian May – *Mission Moon 3-D* – came out of a series of discussions at one of the Starmus Festivals. We had great fun in its wake, in 2018 and 2019, giving lectures here and there, chatting about the history of space-flight, sharing our new stereo imaging of the space race and the Apollo Moon landings with the world, and entangling our love of space and astronaut-explorers. The whole experience was so stimulating we began to wonder about creating another cosmic 3-D book. The work you hold in your hands is the result.

Again, the gang of conspirators was the same. I was elected to write the book, Brian would again be the collaborator and creative director, and our pal Robin Rees was the publisher for the London Stereoscopic Company. What would the theme be this time? We were limited to the entire universe, our first love and what gets us going each and every day. There was a new challenge here. Producing a stereo book of images about the Moon and its exploration is one thing. With the universe at large, however, it's not possible to photograph in true stereo – the stars and galaxies are too far away. But Brian completed our team with one of the truly great astroimagers in the world, Finnish amateur J.-P. Metsävainio, who is a master of simulating 3-D views of deep-space objects. This is the very first time such a collaboration has happened.

Our topic would be the birthplaces of stars, nebulae, the great cosmic clouds that collect gas and dust and elemental goodies and produce new generations of stars as they are drawn down by gravity. Some of the loveliest and most awe-inspiring objects in the heavens are nebulae, and we were enchanted by the idea of telling the story of starbirth and stardeath in a completely new way.

Our friend J.-P.'s images, although they are simulations, are true master-pieces. In creating them, he communicates the subtleties of astrophysics in a brilliant manner. Star clusters in his images, born of the gas, are located centrally; dark nebulae that we see only because they block light from beyond are fore-ground objects; star distances are accurately represented; and he structures a 3-D view of the nebular shells themselves so as to depict them accurately, as precisely as we could hope for. The result is the effect one might get if we were flying through space toward them, and through them.

This subject ties in rather magically with long-standing interests of mine. When I was a teenager, in the untamed wilderness of rural Ohio, I became entranced by the dark sky above. I was given an eight-inch telescope as a present one year and found lots of nebulae, first under cold, crisp winter skies, after having read that you "weren't supposed to see them" in such a small telescope. So empowered was I about observing the universe that I started a small publication, *Deep Sky Monthly*, and that led me to *Astronomy* magazine, where I've been for 37 years.

This is such a happy full circle for me to make – to again visit old friends, the cosmic clouds, that bring us all together as we look toward the sky. We are liter-ally made of the atoms from stars that form from these clouds. How amazing it is that we can know this and read about it and talk about it, as friends on a small planet in a very big universe.

INTRODUCTION

BY BRIAN MAY

Over the years I have learned that great things often come from bringing great people together. I think this book is a good example. This up-to-the-minute exposition of the workings of nebulae, illustrated with highly authentic side-by-side stereographs as well as some of the latest mono nebular astrophotography, is the result of a unique assembly of talent. The original London Stereoscopic (or, initially, "Stereoscope") Company (LSC) was born in 1854, dedicated to purveying high-quality stereoscopy to the public. Stereoscopic photography, alongside photography itself, was new and exploding as a Victorian craze. The LSC's motto was "No home without a Stereoscope," and indeed it came close to achieving that in the 1850s in England, its London shops boasting a million stereo views for sale by 1858. It survived until it was wound up around 1930 – an impressive innings that outlasted two golden eras of stereoscopic photography.

In 2003, with some help from my friends, I began to rebuild the LSC based on nothing but the ashes of the original company and a reborn spirit – belief in the magic of what had by then become known as "3-D." The aim was to bring Victorian stereoscopy, still unsurpassed, renewed using modern techniques, into the 21st century. The LSC is now a small but gradually growing team of people who are all dedicated to excellence in stereoscopy for its own sake. This will be the seventh book we have published. All of our works have been milestones in some way, and this is no exception.

This is the first ever stereoscopically illustrated book dedicated to the study of the vast array of nebulae now known to inhabit our Milky Way Galaxy (plus a couple beyond its limits). Many of these huge clouds of gas and dust are the birthplaces of stars – incandescent orbs which in many cases will end their own lives creating new nebulous areas, in the vast spaces where atoms are constantly being transformed and recycled. Once we realize that the very molecules in our own bodies are part of this huge cosmic drama, it becomes clear that this is something we all need to know about.

Ohio-born David J. Eicher – long-time Editor-in-Chief of *Astronomy* magazine – is an expert in the life and death of nebulae; he was the perfect author of choice to tell this story in words we can all understand. Finnish luminary J.-P. Metsävainio is the only man on Earth to have assembled a whole encyclopedia of stereoscopic images of nebulae, based on his own astrophotography. I'm proud to have brought Eicher and Metsävainio together – two unique talents collaborating for the first time. And, as with all great events, in retrospect, it's apparent that this book couldn't have happened any other way. Behind the text and images is the now experienced team of the LSC, with all the passion for exquisite detail and for breaking new ground that they bring.

The stereoscopic pairs in this book are revolutionary. To create a stereoscopic pair of images which will deliver depth perception in a photographic view, we normally need two separate viewpoints, corresponding to the positions, side-by-side, of our own eyes. If we are capturing a flower close-up in 3-D, the separation of those two viewpoints, known as the "baseline" of the stereo, will be just a centimeter or two (around an inch). This will be enough to give us the slight differences, known as "parallax," between the two images captured, which will enable our eyes and brain to reconstruct the view in true depth when we view this pair of pictures in a stereoscope.

If we are making a stereo portrait of a person, or a landscape with some distinct foreground detail, the baseline can be the same as the actual distance between our eyes (about 2½ inches, or 7 centimeters). But how do we capture true stereo information in objects much farther away? Capturing cityscapes from a high building, we may find that we get an excellent "hyper stereo" effect by extending our baseline to several meters – I have used up to 50 feet to good effect from the observation deck of the World Trade Center building. Moving out into space, we can achieve suitable baselines of effectively hundreds of thousands of miles, simply by waiting for planets and moons to rotate. But if we move to cosmic scales beyond this, it gets tricky. The New Horizons Mission probe, currently over 7,000,000,000 kilometers (about 4,500,000,000 miles) away from the Earth after sensationally passing Pluto and a Kuiper Belt object at close quarters, is JUST far enough away to make it possible to achieve a baseline long enough to see instantaneous parallax in nearby stars. This is an experiment that this innovative NASA team will be attempting in the near future for the first time. But the nebulae we will come to know in this book are beyond the reach of any such technique. They are simply too far away. How, then, can we make informative and beautiful stereos of them? Mr Metsävainio has developed his own unique methods. First of all, he photographs the area in question himself, obtaining a high-resolution color image that necessarily appears "flat" because it is so distant. He then applies current knowledge of the distances of every part of the nebula. From this he constructs a depth map, which he then uses to create a second image in which the required differential parallaxes have been applied, modifying the image laterally. The result is a magical transformation of a flat view into a completely astrophysically informed three–dimensional view. The effect is way beyond astonishing! In fact, these images radically transform our mental appreciation of these objects. What appears in every other astronomical image seen in books as a flat ring of smoke suddenly becomes appreciable as a three–dimensional transparent sphere. Dark patches which look, in flat images, as if they might be holes in the bright parts of the nebula, are now unmistakably revealed to be the result of dark material between us and the illuminated gas and dust behind. In short, these images, never seen before in print, are revelations. It has often been said that a picture is worth 1,000 words. If this is true, a stereoscopic picture must be worth at least 100,000! Luckily in this book we have all three – superlative stereoscopic pictures from Mr Metsävainio, flat pictures from the greatest astrophotographers and telescopes in the world (and just outside!), and a magnificent text from Mr Eicher, one of the world's top experts in the field.

Before you begin reading, please retrieve and deploy your OWL stereoscope, included in the back of this package. If you're not already familiar with stereoscopes, taking a moment to read the instructions may change your life! And certainly take you on a new journey into space.

Enjoy!
Brian May, December 2019.

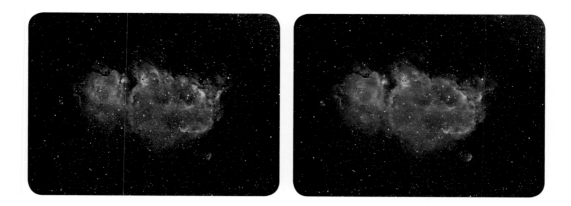

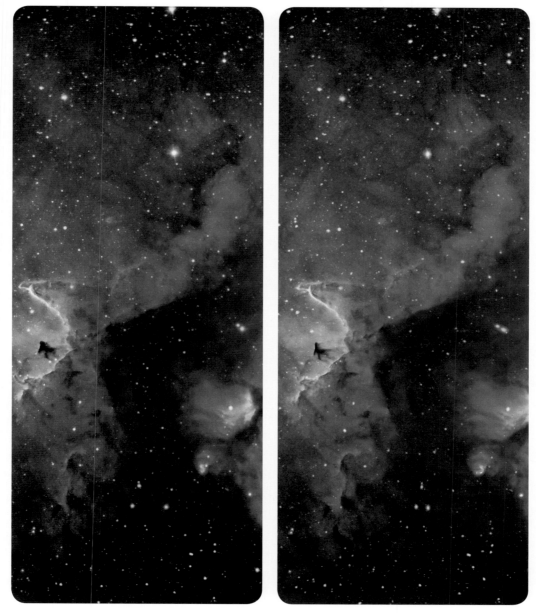

The Soul Nebula in 3-D

The Soul Nebula (Sh2-199, LBN 667) is an emission nebula in the constellation Cassiopeia. IC 1848 is an open star cluster embedded within the nebula. Distance is about 7.500 light-years. This complex is an eastern neighbor of IC 1805, the Heart Nebula, and they are often mentioned together as Heart and Soul. Top is the stereo of the entire nebula and below is a detail view.

Nebulosity around the North Star

Polaris, the North Star, shines brilliantly at the heart of this image of the North Celestial Pole. Galactic "cirrus" clouds of dust billow throughout this region. The open star cluster NGC 188 can be seen just right of center at the bottom of the image.
Credit: Rogelio Bernal Andreo

The Statue of Liberty Nebula

Lying near the Carina Nebula, in the constellation Carina, is the peculiar object NGC 3576. Sometimes called the Statue of Liberty Nebula, NGC 3576 is a towering emission nebula that is collapsing down into newborn stars.
Credit: Don Goldman

CHILDREN OF THE COSMOS

The universe is a magical place. Many people don't see this; they are born, live their lives, and perish, without ever realizing the magic is there. We're not talking about magic in the way you might think – spells, amulets, or any kind of occultism. Over the last century astronomers have discovered that the universe is a place ruled by *natural* magic – nature's own amazing laws that make it a miracle in itself. As more and more of the truth about the heavens – the vast space around us – is revealed every day, we see that it's actually far stranger, and more wondrous, than any of the imaginary stories humans have concocted over the years on our little blue planet Earth.

Consider the very stuff you're made from, for example. The average human has seven octillion atoms in their body. That's ten to the 27th power. Put another way, it's seven billion billion billion atoms. Suffice it to say, it's a lot. These very same atoms were created in the early stages of the universe or in the bellies of exploding stars long ago. As the great astronomer Carl Sagan used to say: "The nitrogen in our DNA, the calcium in our teeth, the iron in our blood, the carbon in our apple pies were made in the interiors of collapsing stars. We are made of starstuff."

Right now, you have at least traces of 60 chemical elements within you. Oxygen is the most abundant by mass; carbon follows second, and then hydrogen and nitrogen. But you also have so-called heavier elements such as calcium, phosphorus, potassium, sulfur, sodium, chlorine, and magnesium. And yes, you even have naturally occurring radioactive elements within you – again, all part of nature.

The elements, of course, are the basic atomic building blocks of the cosmos, from which all normal matter is composed. Consider, just for a moment, our discovery and understanding of them. Organized by their properties in the so-called periodic table, the 118 known elements display a wide range of characteristics. The Russian chemist Dmitri Mendeleev created the first detailed periodic table in 1869, to understand and organize the elements. The first 94 elements occur naturally, and the last 24 have been synthesized in labs or nuclear reactors, but are not yet observed in nature.

The first known elements were metals, dating back to the last part of the Stone Age, when copper was discovered and used as a resource. This took place around 9000 BC, in the Middle East, although early manufactured copper dates from more recent times. Beads found in Çatalhöyük, Anatolia, in what is modern Turkey, suggest the manufacture of copper goods begins about 6000 BC. This is highly interesting, as Çatalhöyük was one of the first and most important proto-city settlements on Earth. Clear evidence of copper smelting appears in 5000 BC, at the archaeological site of Belovode in the Rudnik Mountains of what is now Serbia.

By 6000 BC, humans were also smelting and using lead and gold. Silver and iron came next, before 5000 BC. By the time of the Egyptians, alchemists had discovered carbon, and the smelting of tin by 3500 BC led to the Bronze Age, combining tin with copper to fashion the hardy bronze alloy used in so many ways.

Many of the important elements had to wait until modern times for their discovery. Hydrogen, the most abundant element, was discovered by the English natural philosopher Henry Cavendish in 1766. The Swedish-German chemist Wilhelm Scheele identified a variety of elements during the 1770s, including oxygen, chlorine, manganese, and tungsten. The English natural philosopher Joseph Priestley and French chemist Antoine Lavoisier discovered oxygen at around the same time. The most recently identified element is oganesson (atomic number 118). Unstable and radioactive, it was first synthesized in 2002, and recognized as a new element in 2015.

THE ORIGIN OF THE ELEMENTS

So where did the elements that make up our stars, our planets, and even us, come from? The creation of the first atomic nuclei took place immediately following the Big Bang itself, the origin of the universe some 13.8 billion years ago. That process, called "Big Bang nucleosynthesis," created mostly hydrogen and helium, with trace amounts of other elements like lithium and deuterium.

Long after this, heavier elements were formed in a variety of processes. Stars are nuclear fusion reactors, engines that fuse lighter elements together into heavier ones. The process of fusion itself creates many more elements of the periodic table, up to iron and nickel. Exploding stars, supernovae, are the dying carcasses of massive stars far heavier than the Sun. Their cataclysmic explosions create much heavier elements, and the blasts send them far out into the sur-

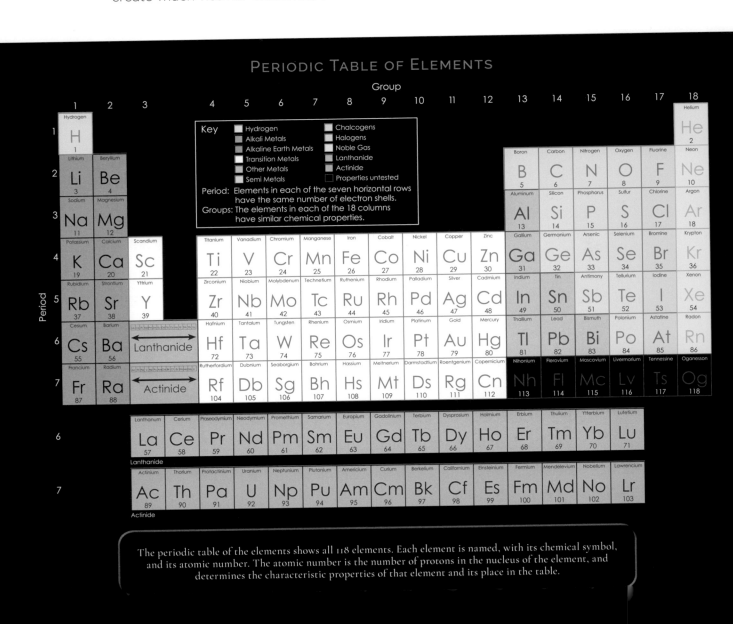

The periodic table of the elements shows all 118 elements. Each element is named, with its chemical symbol, and its atomic number. The atomic number is the number of protons in the nucleus of the element, and determines the characteristic properties of that element and its place in the table.

rounding galaxy. These processes, over vast amounts of time, have created and spread the whole variety of elements that we know of today, including the stuff that makes up our bodies.

The cosmic origins of the elements, then, are varied. About two dozen elements originate from dying low-mass stars. These include carbon, nitrogen, strontium, and tin. Another two dozen or so elements come mostly from exploding massive stars, supernovae. These include oxygen, potassium, sodium, arsenic, and aluminum. Two elements arise from cosmic ray fission; that's the impact of energetic particles from space into Earth's atmosphere and surface. This process creates boron and beryllium.

About another two dozen elements are created largely from merging neutron stars – the clashes of super-dense, dying stellar remnants made mostly of packed neutrons. These include iodine, xenon, cesium, platinum, and gold. And a small number of elements are created, or at least can be created, by exploding white dwarf stars, the final, decayed ultra-dense remnants of stars like the Sun. These include titanium, vanadium, chromium, manganese, iron, and nickel.

These are incredible facts to ponder as you walk out under a starry sky on a clear, moonless night. Look deep toward the shimmering glow of the Milky Way, and you'll see many twinkling stars and the unresolved light from millions more that make up the hazy band running across our sky. That oldest of all human questions, "Why am I here?" now has some kind of an answer. You're here because atoms created in the Big Bang and in stars have recombined, billions of years after their creation, in a way to make your body – with a big thank-you to your parents as well! But perhaps this is actually the answer to "*How* did we get here?" The question of "Why?" perhaps still needs to be resolved.

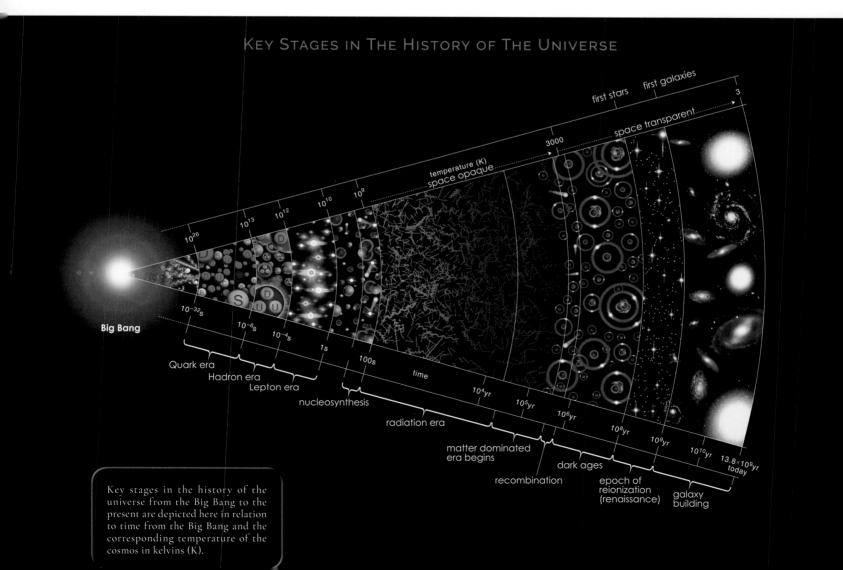

Key stages in the history of the universe from the Big Bang to the present are depicted here in relation to time from the Big Bang and the corresponding temperature of the cosmos in kelvins (K).

THE LIVES OF STARS

The story of elements in nature, of how we got here, of our cosmic roots, is strongly tied to the story of stars in our galaxy and universe. And to tell it we need to explore the lives of stars, how stars come to be, what happens during their lifetimes, and how they, too – like humans – eventually die. We may not all know it, but we are part of the biggest recycling program that exists – the birth, life, and death of stars.

Stars are born in great clouds of gas and dust called "nebulae," or "interstellar clouds." The word nebula comes from the Latin, and means "cloud" or "fog." The gases are mostly hydrogen and helium, and they are typically ionized. That means they are excited, energized, by the radiation from hot stars inside and close to them, and that causes them to glow – to light up. This is lucky for us, because it enables us to see them at very large distances across our galaxy, or even in other nearby galaxies. Astronomers believe the dust in the nebulae is the remains of supernova explosions.

Recent observations of the universe have shown that not only is the cosmos expanding, as we have known for more than a century, but that the universal expansion is accelerating over time. On large scales, everything is moving away from everything else, and the universe is getting bigger.

But various forces are at work in the universe. One of the most important, the very force that keeps us on Earth's surface, is gravity. The attraction of gravity dictates that, even though the universe is expanding, things that are near each other are drawn together because of their mass. This means that galaxies close to each other will often merge together as one. It also initiates the birth of new stars, as gravity causes the gas and dust in nebulae to be condensed down into smaller volumes. As this process occurs, eventually enough mass accumulates to form a new star – hydrogen, helium, and other elements are compressed until a critical mass and density is reached, and a new nuclear fusion reactor – a star – is born.

So, the majority of nebulae we see scattered across our night sky almost all belonging to our Milky Way Galaxy, are cauldrons that make possible the births of new stars. They are stellar nurseries, and that's why they are frequently intermingled with clusters of young stars. By observing nebulae, we are peering into the world of infant suns, seeing a process that for our own Sun took place some 4.6 billion years ago.

This is an important narrative to understand and appreciate, because it truly allows us to see where we came from, and why we're here on a planet orbiting one rather ordinary star in the Milky Way. To understand the universe is paramount for the millions alive today, and also for the unborn millions to come. This is a detailed story, and we aim to share all of it with you in this unique book. We'll learn about many aspects of the universe as we unravel the tale of cosmic clouds. It involves detours into science, history, and maybe a bit of philosophy. It's a subject that is easy to become passionate about. We're reminded of a quotation attributed to Abraham Lincoln: "As the preacher said, I could write shorter sermons, but once I get started, I get too lazy to stop."

The majority of nebulae in our sky are stellar nurseries. But not all of them. We will survey other types of gas clouds, too. Some are simply blobs of gas that are not energized and glowing of their own accord but are reflecting the light from bright stars along our line of sight. Others are truly dark nebulae, composed of dusty, black grains, obstructing the light from stars beyond them. Thus, we

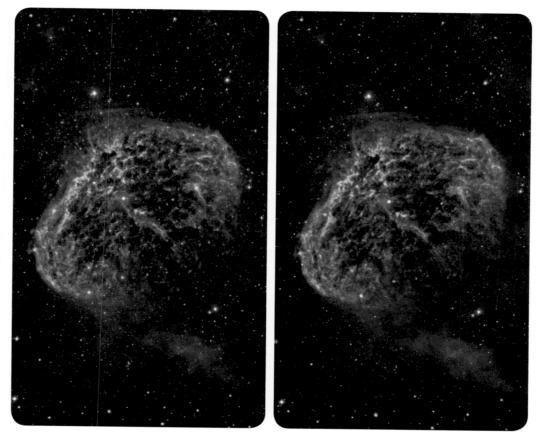

The Crescent Nebula in 3-D

A stereo view of the Crescent Nebula (NGC 6888) shows this object as an oblate bubble of gas encapsulating its hot central star, with blobs of multicolored nebulosity aligning here and there to form the three-dimensional cloud.

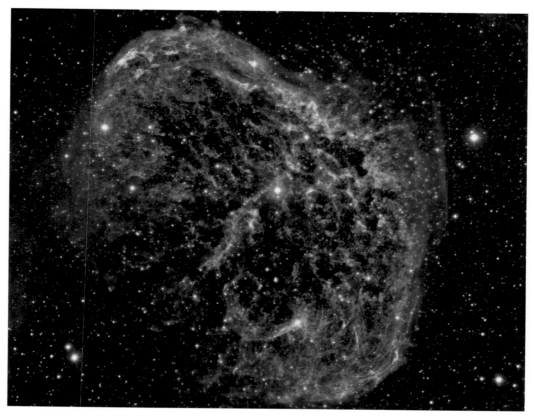

The Crescent Nebula

The Crescent Nebula (NGC 6888) is a favorite summertime object for observers, although it is relatively faint for optical telescopes. Discovered by William Herschel in 1792, this strange object is a glowing cloud formed when a high-energy star flung away high-velocity stellar winds. The star is a so-called Wolf-Rayet star, a rare type that shows strong stellar winds, and the high-velocity material is colliding with older, slower-moving gas ejected when the star became a red giant, hundreds of thousands of years ago. *Credit: J.-P. Metsävainio*

The Wizard Nebula in 3-D

In stereo, the Wizard Nebula (NGC 7380) appears as a wall of bright and dark nebulae, arcing steeply upward, with scattered foreground stars and wisps of closer dark nebulae peppering the view.

see them in outline, as ghostly clouds of blackness, floating in the immense void of space. Some other glowing nebulae are the remnants, the torn-away insides, of massive stars that have violently exploded as a cosmic bomb. These leavings glow for a short cosmic time before dissipating into invisibility. Still other types of nebulae are the endpoints of ordinary stars like the Sun, shells of softly glowing gas that cocoon outward, belched away by the dying carcasses of their progenitor stars within. Each of these types of nebulae offers numerous varieties and examples, and we will survey many of them, in both stereo and mono imagery, for your visual pleasure.

Exploring the world of nebulae offers an eye-opening understanding of the cosmos at large. This is especially true if we keep chemistry in our minds as a theme. We're able to understand a great deal about the universe because of chemistry. Specifically, spectroscopy (see chapter 3) is a vital and powerful tool for astrophysicists. By carefully analyzing spectra, showing variations in the intensity of colours, in the light from various astronomical objects, astronomers can understand the chemistry of the objects they're looking at. And countless millions of spectral observations of stars, planets, galaxies, and so on have demonstrated that chemistry is uniform throughout the cosmos. That is, it works the same way in a galaxy ten billion light-years away as it does in your backyard. (A light-year is the distance light travels in one year, about six trillion miles, or 9.5 trillion kilometers.) And that's a crucially important fact that astronomers use to understand how the universe works.

Chemistry is everywhere in the cosmos. All matter that exists in the universe is made of chemicals. The only thing we experience every day that's not made of chemicals is a thought – but our thoughts are themselves byproducts of chemical interactions within the brain.

Let's step away from ourselves for a moment to consider how matter comes together. Again, there's no magic involved in the way the universe assembles things. Consider some of the most abundant objects on our planet, rocks and minerals. Being very interested in planetary geology, one of us (DE) has in

his house a collection of about 2,000 mineral specimens. These are beautiful, showy crystals, and they're exciting because they show the way the universe builds planets. There's no randomness here, nor any magical behind-the-scenes thought or preordained control. A simple pyrite crystal builds itself when iron and sulfur atoms are in solution in the right abundances. The atoms are electro-chemically attracted to each other, and they assemble in a crystal lattice in just the right way to create a crystal. The more solution that's available, the larger the crystal can grow. The same is true of the whole spectrum of about 5,400 known minerals, including emeralds, diamonds, quartz, gold, garnets, wulfenite, rhodochrosite, and many others.

Holding a mineral specimen in your hand can be a special experience, because of what spectroscopy tells us. From that technique, as stated, we know that chemistry and physics are consistent throughout the universe. Temperatures, pressures, and many other local conditions could be wildly variable, but count-less other worlds throughout our galaxy and the universe might contain miner-als very much like the ones we have on Earth. So mineral specimens give us a window into far-away worlds that we will never see up close.

That's intriguing because the more we look around our area in the Milky Way galaxy, the more we have discovered that many other planetary systems exist around nearby stars. Our galaxy consists of several major parts, but the most recognizable and distinctive is the Milky Way's disk, the brightest portion where most of the stars, gas, and dust exist. The Milky Way contains some 400 billion stars spread across this flattened disk, some 100,000 light-years across. Our Sun, of course, is just one of the roughly 400 billion stars.

We've known something about our planetary system since prehistory, of course, when ancient skywatchers named the naked-eye planets after gods because they had the power to move night to night relative to the fixed stars. We've been all the way through the discovery of Pluto in 1930 and its demotion to dwarf planet in 2006, and we're now aware of the huge population of smaller bodies in our solar system, dwarf planets, so-called Kuiper Belt objects, comets, and asteroids. They are almost countless, and many thousands are cataloged and named. That's without including the billions of the smallest members of the Sun's family ... the particles in the Zodiacal Dust cloud that circle roughly in the plane of the planets.

ARE WE ALONE?

Only in recent times have astronomers had the power to discover planets orbiting stars other than the Sun. Technological advances in telescopes and observing methods brought the first confirmed discovery of a so-called exoplanet – shorthand for extrasolar planet – in 1992. As of the summer of 2019, we now know of nearly 4,100 exoplanets in more than 3,000 systems, and astronomers have only reached out to relatively nearby space in our galaxy.

Because of the difficulty of detecting planets orbiting stars from enormous distances, many of these planets are massive – so-called "hot Jupiters" that are relatively close to their suns. The most productive planet hunting instrument was the Kepler Space Telescope, which orbited Earth and cataloged exoplanet discoveries from 2009 through 2016. This magnificent telescope studied a rela-tively small area of sky and found more than 2,600 of the nearly 4,100 known

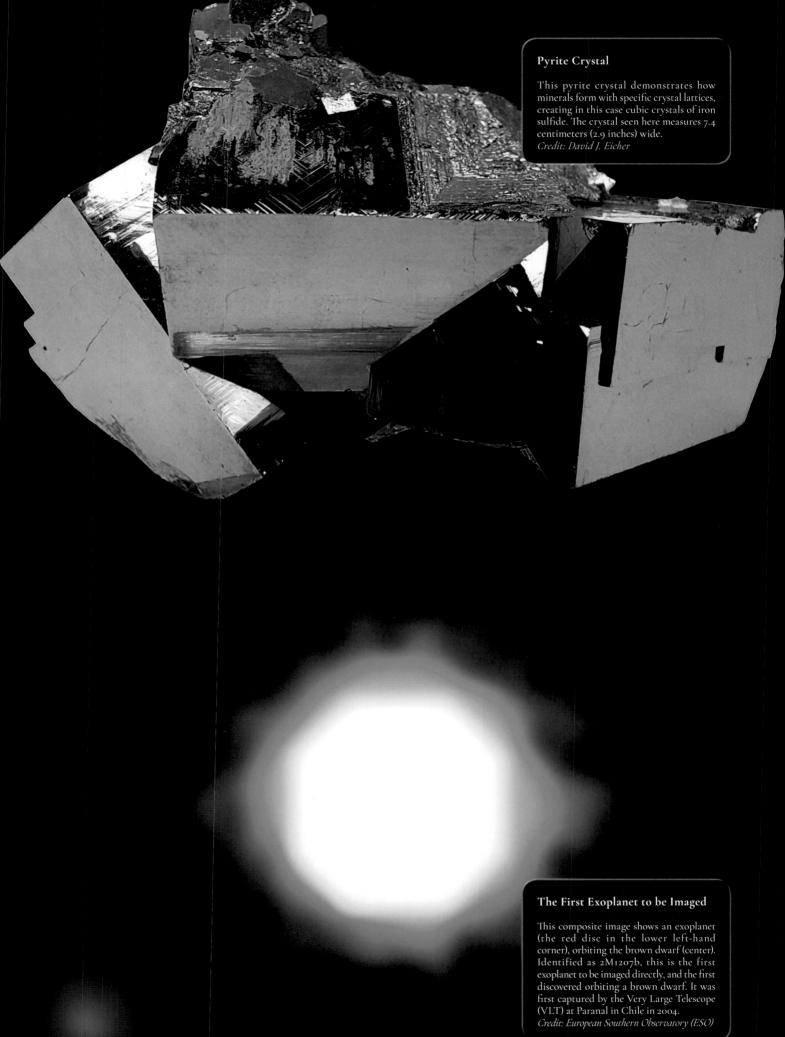

Pyrite Crystal

This pyrite crystal demonstrates how minerals form with specific crystal lattices, creating in this case cubic crystals of iron sulfide. The crystal seen here measures 7.4 centimeters (2.9 inches) wide.
Credit: David J. Eicher

The First Exoplanet to be Imaged

This composite image shows an exoplanet (the red disc in the lower left-hand corner), orbiting the brown dwarf (center). Identified as 2M1207b, this is the first exoplanet to be imaged directly, and the first discovered orbiting a brown dwarf. It was first captured by the Very Large Telescope (VLT) at Paranal in Chile in 2004.
Credit: European Southern Observatory (ESO)

The Hubble Deep Field

The estimate of 100 billion galaxies in the universe is based on this Hubble Deep Field image of a small region in the constellation of Ursa Major. Covering an area of about 2.6 arcminutes (about 24 millionths the area of the whole sky), it was made from 342 separate exposures captured by the Hubble Space Telescope. *Credit: NASA*

exoplanets. A newer telescope, the Transiting Exoplanet Survey Satellite (TESS), was launched in 2018 and has begun another episode of exoplanet detection.

As we look out into the galaxy surrounding our solar system, it's not surprising to see lots of nearby planetary systems. Astrophysicists believe that stars form as cosmic clouds, and these nebular stellar nurseries collapse and stars wink on inside them. And the leftover detritus from the collapse, swirling slowly around the infant suns, makes a cadre of planets and smaller bodies surrounding the new star.

So, it all seems to make sense, and we are in the pioneering days of being able to see farther out into the galactic neighborhood that surrounds us. Understanding that stars are so numerous and that planetary systems are plentiful is exciting. After all, the most basic driving question everyone would like to answer is at the base of it all: Are we alone? Is there other life in the solar system? In the Milky Way? In other galaxies?

We just don't yet know. The numbers are staggering. By taking very deep exposures of small areas of sky with the Hubble Space Telescope, astronomers have estimated that something like 100 billion galaxies must exist in the universe. And that's just in the *visible* universe, which almost certainly does not represent the whole universe that exists. But for simplicity's sake, let's say that it does. Let's say that an average galaxy contains 100 billion stars, as many dwarf galaxies are smaller than our galaxy. Multiplying that out gives us an approximate number of stars in the universe as something like 10,000 billion billion. That's an awful lot of stars – and an awful lot of planets.

Is it possible that our little blue planet Earth is the only place in the entire universe with life? Or with a civilization? It would seem hard to believe. Our whole heritage of discovery in astronomy commenced with Earth being in the center of everything, the most special place there is, and subsequently we found out how disastrously wrong that idea was. And we know that chemistry is uniform throughout the cosmos, and that complex organics, the stuff of life, exist in all manner of places out in space. And yet we know, thus far, of just one planet in the cosmos that hosts life – ours.

Meanwhile, this story of nebulae – cosmic clouds – can shed much light on how we came to be here on this one small watery world, and what the universe holds at large. As Stephen Hawking said, as humans, "We need to know." After all, the very building blocks of our bodies were born in the Big Bang, and in the explosions of massive stars. We are indeed children of the cosmos.

Planet Earth

Stereo view of our blue planet made from photographs captured by Apollo 8.
Credit: NASA/Brian May

Reflection Nebula in Perseus

The prominent reflection nebula
NGC 1333 lies about 1,000 light-years
distant and contains a large number
of brown dwarfs, low-mass objects
that are not quite massive enough to
start nuclear reactions as stars.
*Credit: Adam Block/Mount Lemmon
SkyCenter/University of Arizona*

The Milky Way

The sky is filled with nebulae glowing mostly in red or blue light. A wide-angle view of the Milky Way made at a California star party shows numerous bright and dark nebulae intertwined with the plane of the Milky Way, shown here as it appears in summer evenings in the northern hemisphere. The view encompasses Scorpius near the horizon, northward, toward Cygnus near the top.
Credit: Tony Hallas

A TRIP THROUGH THE MILKY WAY

In our rapidly growing, somewhat careless 21st century world, it's not always easy to see the sky at night. But if you can drag yourself away from a city, and the attendant light pollution it creates, get to a dark sky site. On a moonless night, look skyward, and throughout the night, you'll see the glowing band of the Milky Way arcing throughout the sky. The best way to view the Milky Way is in the wintertime and summertime evening skies.

Among the highlights of the winter Milky Way at northern latitudes are the brilliant constellation Orion the Hunter, with its famous "belt" and human-like shape, and the bright Orion Nebula nestled in the hunter's sword. The summer Milky Way features splendid treats as we look toward the center of our galaxy, which lies in the direction of the constellation Sagittarius. Countless star clusters and nebulae lie scattered across this rich region, and the neighboring constellations Scorpius and Centaurus.

The name Milky Way derives from the Latin *via lactea*, which in turn comes from the older Greek term γαλαξίας κύκλος, *galaxías kýklos*, "milky circle." The Milky Way we see as a band of light running through our sky comes from the unresolved light of millions of faint stars that combine to create this ethereal glow. The story of discovering the true nature of our galaxy was a long time in coming, in part because it was difficult to determine the nature of the universe, and in part because it's hard to decipher the structure of our galaxy when we are inside it.

For hundreds of years, astronomers faintly perceived that the universe was a large place, and that lots of stars inhabited the sky, but they really had no idea about the extent of the cosmos. In the 16th century, Copernicus opened our knowledge to a Sun-centered (heliocentric) cosmos (rather than placing Earth in the center of it all), and a century later Galileo studied the sky with his telescope and determined that the Milky Way is made of innumerable stars. Astronomers armed with telescopes studied many

Orion Rising

In this image taken from California, Orion rises just over the horizon, and the Milky Way traces northward. Visible above and left of center are the Hyades cluster with the bright star Aldebaran and the Pleiades cluster above. *Credit: Tony Hallas*

of the fuzzy clouds of light in the sky, and by the 19th century Irish astronomer William Parsons made drawings of "spiral nebulae" that showed a clear pattern of apparent rotation. But no one knew what they were, and it would be 1923 before Edwin Hubble could discover that spiral nebulae are enormous galaxies, systems of stars and gas, and that they were far more distant than anyone had imagined before the 20th century.

We now know that the Milky Way Galaxy, in which our Sun is just one star, contains something on the order of 400 billion stars. The exact number is hard to pin down because the most numerous stars, red dwarfs, are faint and hard to detect over large distances. The discovery of the nature of our galaxy is quite recent. We have no images of the Milky Way from the outside, of course, and much of what we can see inside our galaxy is obscured by dust – dark nebulae – making piecing together a coherent picture of our galaxy's form somewhat challenging.

For decades, astronomers believed the Milky Way was an ordinary spiral galaxy, with a central hub, spiral arms, and an outer halo of material. The brightest spiral galaxy in our sky, the Andromeda Galaxy, shows that familiar form in images, and we believed the Milky Way to be similar in shape. But in 2008 a team using the Spitzer Space Telescope, a Sun-orbiting observatory, found that our galaxy has a barred spiral structure, with a prominent bar-shaped wall of stars and gas across its center. Our galaxy's spiral arms originate from the ends of this bar.

We now have a pretty fair understanding of both the shape and the composition of our Milky Way. Our galaxy is important to the study of nebulae, as nearly all of these objects we can see and study are within the Milky Way. Only a small number of nebulae in other very nearby galaxies are visible to us because of the great distances to other galaxies. So, let's consider the basic parts of our galaxy, and the relationship of nebulae to our galaxy as a whole.

Most of the stars within our galaxy lie within the Milky Way's bright disk, a thin "plate" of stars, slowly rotating around the galaxy's center, that also contains

A Dark Nebula in Orion

B32, the dark notch at center, juts in front of the strip of bright nebulosity cataloged as Sharpless 2–264 in Orion. Part of a huge circular complex of nebulosity known as the Lambda Orionis Ring, this nebular cloud offers a good look at an opaque dark cloud. The bright nebulosity lies about 1,400 light-years away and B32 is a foreground object. *Credit: Alistair Symon*

much of the galaxy's gas and dust. It's a portion of the galaxy's disk that we see in our sky as the band of milky light. The overall diameter of the Milky Way's disk is thought to be at least 100,000 light-years, although recent studies of the last couple of years suggest it may be substantially larger, perhaps even 200,000 light-years across. Imagine the galaxy's disk as a huge, slowly spinning CD rotating around the galactic center.

Within the galaxy's disk, the Milky Way's central bar extends some 11,000 light-years from the center. The Milky Way has at least six principal spiral arms, extending outward: the so-called 3-kpc Arm and Perseus Arm, the Norma and Outer Arm, the Scutum-Centaurus Arm, the Carina-Sagittarius Arm, and two smaller arms or spurs. One of the two smaller arms is the Orion-Cygnus Arm, which contains the Sun and the solar system. Our Sun lies about 26,000 light-years out from the galaxy's center, in the relatively quiet galactic suburbs. It's a good thing, too, because multitudinous dangers exist closer to the center – a supermassive black hole, antimatter, colliding stars and gas clouds, and all manner of fun.

The galaxy's slowly rotating disk actually consists of two components, a thin disk and a thick disk. The thin disk contains most of the stars in the galaxy, some 90 percent, and all of the young, massive stars that are being born in open star clusters. Some 4.6 billion years ago, the Sun formed in one of these clusters, and the tidal forces over time, as we have orbited the center of the galaxy, have dispersed all of the Sun's brethren, so that we are left with a solitary star. The arms also contain the glowing nebulae, the stellar birthplaces and endpoints of a star's life, that form the basis of our book.

Our rather remote location, way out from the galaxy's center, means that we make one complete orbit around the galaxy about once every 220 million years, moving at some 240 kilometers (150 miles) per second. So in the course of the Sun's lifetime, we have spun around the galaxy's disk about 20 times. The thin

Sketch of M51

The first sketch of a spiral "nebula" (galaxy), as published in 1850 by William Parsons (the Third Earl of Rosse). His 72-inch telescope, built in 1845 and colloquially known as the "Leviathan of Parsonstown," was the world's largest telescope, in terms of aperture size, until the early 20th century.

The Andromeda Galaxy

For many years astronomers believed the Milky Way was an analog to our sister galaxy in the Local Group, the Andromeda Galaxy, shown here in mono and stereo. Now we know that the Milky Way is a barred spiral rather than a straightforward spiral like Andromeda. Shown along with the Andromeda Galaxy here are its two closest satellite galaxies, M32 (above) and NGC 205 (below).

Credit: J.-P. Metsävainio

NGC 4302

Seen edge on, the galaxy NGC 4302 demonstrates tremendous similarities with our own Milky Way Galaxy, seen below. Captured by the Hubble Space Telescope, huge swathes of dust are responsible for the mottled brown patterns, but a burst of blue to the left side of the galaxy indicates a region of extremely vigorous star formation.
Credit: NASA, ESA, and M. Mutchler (STScI)

Milky Way Panorama

This 360-degree high-resolution image of the Milky Way was created by Serge Brunier at the European Southern Observatory. The plane of our galaxy is seen edge on from the perspective of Earth and cuts through the center of the image; the view is almost as if we were looking at our galaxy from the outside. From this vantage point, the general components of our spiral galaxy come clearly into view, including its disk, marbled with both dark and glowing nebulae, which harbour bright, young stars, as well as the galaxy's central bulge and its satellite galaxies. As filming extended over several months, objects from the Solar System came and went through the star fields, with bright planets such as Venus and Jupiter leaving their traces.
Credit: S. Brunier/ESO

disk that we live in is some 1,500 light-years thick. Encapsulating this thin disk is the thick disk, which is some 3,000 light-years high, and contains an older set of fewer stars that formed earlier in the galaxy's history. Most of the gas and dust in the galaxy – including all of our treasured nebulae – exists in a band no more than 500 light-years thick.

If we traveled toward the center of our galaxy, we would be moving in from a distance of about 26,000 light-years, out here in the suburbs. Looking toward the center of our galaxy, to the constellation Sagittarius, we can see only about a fifth of the way toward the galactic center. That's because this direction is crowded with thick clouds of obscuring dust, and numerous stars. If we approached the center of the galaxy, we would see the density of stars increase dramatically, with a fog of gas and dust.

The galaxy's center itself holds a supermassive black hole, a region of space where the gravity is so intense that nothing, not even light, can escape it. Called Sagittarius A* (pronounced Sagittarius A-star), our black hole has the mass of about 4.3 billion suns. And all of that mass is squeezed into a sphere about the size of Mercury's orbit around the Sun.

The Hercules Cluster

M13, the Hercules Cluster, is located in the constellation of Hercules at a distance of 25,000 light-years. It is one of the brightest globular clusters in the sky.

The galaxy also has a central bulge, around the black hole and galactic center, and contains some 20 billion solar masses of material within it. This central bulge shines with about five billion times the light of the Sun. And far away from the center, the bar, the thick and thin disks, is the galaxy's halo. This outer element contains metal-poor, old stars arranged in what are called globular star clusters, spheres of yellowish suns. You can see many of these in backyard telescopes, and leading examples of them include Omega Centauri and the Hercules Cluster. The halo also contains vast amounts of so-called neutral hydrogen gas, consisting of hydrogen atoms with one proton and one electron, and large amounts of dark matter. This halo reaches at least to 200,000 light-years on either side of the galaxy's center.

This enormous galaxy, our Milky Way, is like a stellar playground for enthusiasts who want to explore the cosmos. Aside from the several hundred billion stars, our galaxy contains several thousand glowing nebulae, and hundreds of star clusters, many of them associated with nebulae.

Consider our own astronomical backyard for just a moment. Something we humans on Earth fail to appreciate is the incredibly vast scale of the cosmos, of even the nearest objects to us. Think of the solar system, our Sun and planets, as a model built on one centimeter representing the distance between Earth and the Sun. (This is called one Astronomical Unit, and in reality equals about 150 million kilometers, or 93 million miles.) On this scale, the farthest distance we humans have traveled – to the Moon – is but a tiny fraction of that one centimeter. On that scale, you couldn't even see it clearly.

On this scale, Mars would be a little farther out than Earth. Jupiter would be five centimeters from the Sun, and Saturn nine centimeters. Uranus would typically be just shy of 20 centimeters out, and Neptune in the region of 30 centimeters, with Pluto some 40 centimeters away. But the edge of our solar system, the Oort Cloud, would be some ten football fields away, end to end. And that's just the physical edge of our own solar system, about one light-year in reality.

Aside from the Sun, the closest star is Proxima Centauri, which lies in a system of three stars, some 4.2 light-years away. Located in the southern sky, light from this star takes 4.2 years to reach your eyes. And the photons you can see from this star are moving at 300,000 kilometers per second (186,000 miles per second). That's the fastest speed in the universe, and it's only possible because the photons have no mass.

Space, as Douglas Adams wrote, "is big. Really big. You just won't believe how vastly hugely mindbogglingly big it is. I mean you may think it's a long way down the road to the chemist, but that's just peanuts to space." And he was right.

Let's take a tour of some of the most prominent nebulae in our sky, starting with a gaze toward the center of the galaxy, the most richly packed area of cosmic clouds, gas, and dust. The galactic center lies in the direction of the constellation Sagittarius. Moving outward, but relatively nearby, we encounter dark nebulae floating in the foreground, and blocking light from stars and gas beyond.

The main line of the bright nebulae scattered throughout Sagittarius and Serpens, along the arch of the Milky Way in our sky, offers some of the best and brightest nebulae we can see. They include the Lagoon Nebula (M8), which lies 4,000 light-years away; the nearby Trifid Nebula (M20), also about 4,000 light-years away; the Omega Nebula (M17), some 5,000 light-years distant; and the Eagle Nebula (M16), which lies at a distance of 7,000 light-years.

A word about distances to nebulae: it's difficult for astronomers to measure the distance to a glowing cloud of gas. Fortunately, nearly all nebulae are involved with either newborn or dying stars, and so the distances are generally derived from the stars themselves. That said, the distances to some nebulae are not yet known to a high precision. But we do know that in the case of looking at the rich areas of nebulae toward the center of the Milky Way, we are seeing but a short distance. Vast clouds of dust in the Sagittarius nebulae block our view of the busiest areas of the Milky Way. They are mostly situated about one-fifth of the way toward the core of our galaxy.

If we look in the opposite direction of the galactic center, we can also see lots of beautiful nebulae inhabiting the winter Milky Way, and existing in outward spiral arms of our galaxy. This brings into play such winter constellations in the northern hemisphere as Orion, Taurus, Perseus, Gemini, and Monoceros. One of the closest star clusters to us is the famous Pleiades (M45) in Taurus, often called the Seven Sisters, as sharp-eyed viewers can spy seven stars in this dipper-shaped group with just their naked eyes. One of the stars, Merope, contains some fairly bright reflection nebulosity, which is designated NGC 1435. This dusty area is a challenge to see with telescopes but the nebulosity can be easily recorded by astroimagers. The Pleiades and its associated nebulosity lie some 445 light-years away.

The Pleiades Cluster

The Pleiades star cluster in Taurus, traditionally known as the Seven Sisters for its gleaming, tightly-packed, naked-eye stars, is wrapped in a cloud of faint reflection nebulosity. The brightest portion is the nebula NGC 1435, which surrounds the star Merope, below center.
Credit: Terry Hancock

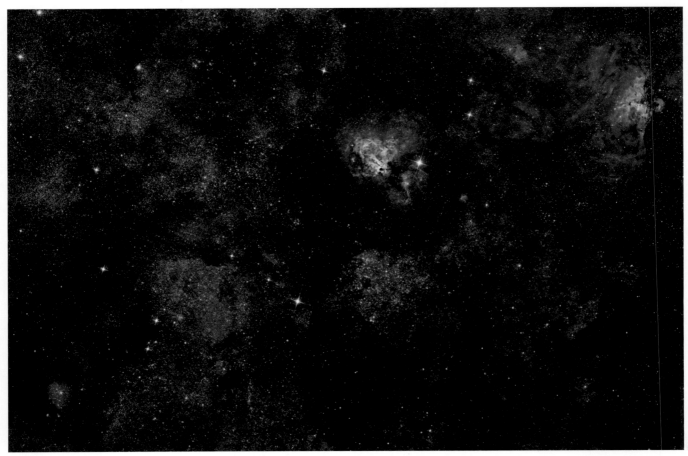

Nebulae and Clusters in Sagittarius and Serpens

An exceptionally rich area of Milky Way in Sagittarius and Serpens contains a multitude of nebulae and clusters. They include the Eagle Nebula (M16, near top right), the Omega Nebula (M17, above and right of center), nebula IC 4701 (below M17), and nebula IC 1283/4 (near bottom left). The star cloud M24 lies to the right of IC 1283/4, and the bright cluster M25 lies at the top left edge. *Credit: Terry Hancock*

Stepping farther out away from us, toward one edge of our spiral galaxy's disk, we encounter lots of winter's brightest nebulae. They include the California Nebula (NGC 1499) in Perseus (see page 102), so named for its resemblance to the state, a glowing star-forming region that lies 1,000 light-years away. Nearby, the constellation Orion the Hunter gleams brightly as the showpiece of the winter Milky Way.

The brightest "hot spot" in all this stellar formation is the Orion Nebula (M42) itself, one of the brightest gas clouds in the entire sky, and probably the most famous of all. Visible to the naked eye as a misty spot, in the sword of Orion (see chapter opening), the nebula is churning out new stars as its gas coalesces by gravity. A telescope shows a wonderland of stars and gas in this region, and the nebula lies at a distance of 1,350 light-years from us. We discuss the Orion Nebula in more detail in the next chapter.

About twice as far away, at a distance of 2,700 light-years, lies another celebrated object, the Cone Nebula (NGC 2264) and its associated star cluster, in the dim constellation Monoceros (see pages 92-3). The dark, cone-shaped nebula associated with this star cluster, called the Christmas Tree Cluster for its distinctive shape, is one of the winter sky's more easily recognized dark nebulae.

Still farther away lies an impressive supernova remnant in Gemini, IC 443. Lying in a rich star field and consisting of a strongly curved arc, this nebula is sometimes called the Jellyfish Nebula (see page 132). It is the remains of a supernova that exploded perhaps 10,000 years ago, and lies about 5,000 light-years away. The number of observable supernova remnants in the sky is relatively small, and this is one of the better targets for backyard observers (see chapter 6).

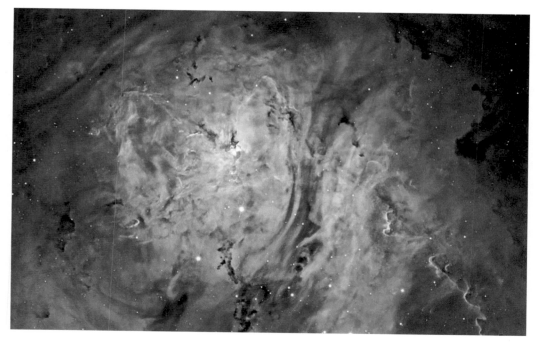

The Lagoon Nebula

The Lagoon Nebula, M8, is a giant interstellar cloud in the constellation Sagittarius. It is classified as an emission nebula and as an H II region. It was discovered by Giovanni Battista Hodierna before 1654 and is one of only two star-forming nebulae faintly visible to the eye from mid-northern latitudes. Hodierna, like Charles Messier after him was cataloging objects he observed in the sky in order that they should not be confused with comets. *Credit: Don Goldman*

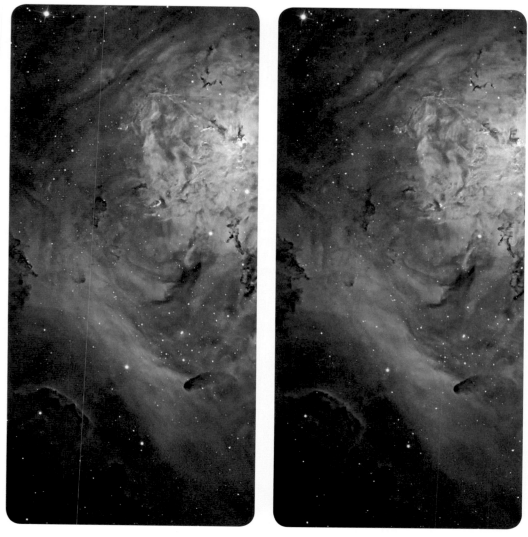

The Lagoon Nebula in 3-D

A stereo view of the Lagoon Nebula shows it as bright nebulosity buried within a cave that encloses a shell of fainter gas and appliqués of dark nebulosity stretching into the foreground.

The Rosette Nebula

One of the most spectacular star-forming regions in the winter sky is the Rosette Nebula (NGC 2237–9), about 5,200 light-years away in Monoceros. This incredible object is faint, and requires a dark sky to be seen visually, but can be photographed relatively easily. Its enormous wreath shape stretches over an area much larger than that of the Full Moon. The central "hole" in the nebula has been cleared out by fierce stellar winds from the young, hot stars within, created by the collapsing gas. The central star cluster is NGC 2244. Areas of the nebula show intricate patterns of dark nebulosity, columns of dust gravitationally collapsing and helping to form new stars.
Credit: J.-P. Metsävainio

Other relatively distant targets oriented away from the galactic center include the large and impressive Rosette Nebula (NGC 2237-9) in Monoceros. This wreath-shaped stellar nursery lies some 5,200 light-years distant, and contains a very bright star cluster in its center, NGC 2244. It can be challenging to observe in backyard telescopes in part because of its large size, some twice the diameter of the Full Moon on our sky. At its distance, then, the Rosette is one of the larger star-forming regions we know of in the galaxy.

Possibly the most famous supernova remnant also lies in this general direction, the Crab Nebula (M1) in Taurus (see chapter 3).

Hundreds of nebulae lie scattered throughout the rest of the sky, over a great range of distances. They include one of the most prominent dark nebulae, the Coalsack Nebula, in the southern constellation Crux, the Southern Cross. This inky black spot, easily visible to the naked eye, lies some 600 light-years away. One of the closest planetary nebulae, the Helix Nebula (NGC 7293) (see chapter 7), can be found in the constellation Aquarius, lying at a distance of 650 light-years. A very faint and large supernova remnant, the Vela Supernova Remnant, lies in the deep southern sky at a distance of 800 light-years (see chapter 6).

Still more distant objects offer a great variety of sights. In the northern constellation of Vulpecula lies the curious Dumbbell Nebula (M27) (see chapter 3), named for its distinctive shape, one of the best planetary nebulae in our sky. And not far away in Cygnus is the magnificent Veil Nebula (NGC 6960, NGC 6979, and NGC 6992-5), one of the best supernova remnants in the sky. This huge wreath, of which we see glowing portions of a big shell, lies 1,500 light-years away (see pages 75-77 and 136-140).

Such a range of objects in our galaxy and the universe reminds us of what an incredible place we live in, and foreshadows the stories of meaning, of origin, of fate, and of substance, that nebulae can tell us about why we're here on planet Earth.

The Butterfly Nebula

The Butterfly Nebula (NGC 2346) is a beautiful planetary nebula in the constellation Monoceros, seen in the wintertime evening sky at northern latitudes. The winged shape of this object is visible in backyard telescopes, and the central star is unusually cool for such an object. The star is a binary, and it's possible that dust around the star makes it somewhat variable in light output and also helped to shape the butterfly form of the nebula. The whole assemblage lies some 2,000 light-years away. *Credit: Adam Block/Mount Lemmon SkyCenter/University of Arizona*

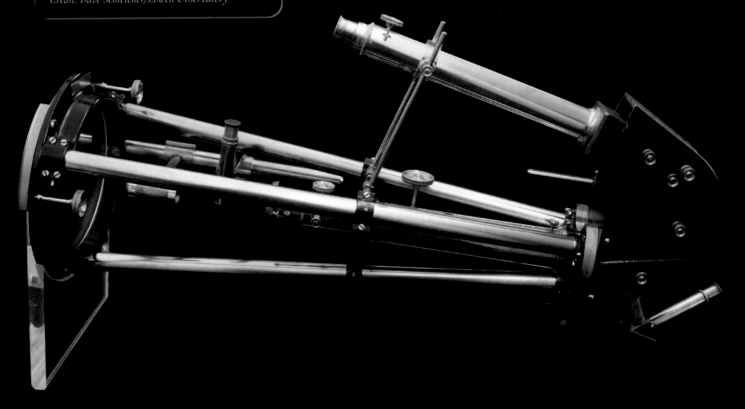

THE INTERSTELLAR MEDIUM

In 1912, at the Lowell Observatory in Flagstaff, Arizona, a young astronomer from Indiana was busily working on several of his pet projects. Vesto M. Slipher had been brought west by the observatory's founder, Percival Lowell, in 1901. Now, in the year of the sinking of the *Titanic*, Slipher was obsessed with spectroscopy. This relatively new tool of astrophysics was transforming astronomy from a science of categorizing and classifying to one of understanding the physical properties of objects in the universe.

Shortly after his arrival at Lowell, Slipher oversaw the implementation of a newly-constructed spectrograph, made by John A. Brashear of the Allegheny

The Pleiades Cluster in 3-D

This stereo image shows how the Pleiades cluster is surrounded by blue reflection nebulae.

Observatory in Pittsburgh. The instrument would allow astronomers to analyze starlight, and light from other celestial objects, thereby understanding some of their physical properties. Slipher mounted it on the 24-inch Clark refractor at Lowell and began a series of studies. He used this instrument to discover that so-called spiral nebulae mostly all seemed to be moving away from each other at high speeds, thereby revealing that the universe is expanding. A decade later, Hubble would demonstrate that spiral nebulae were in fact galaxies.

But Slipher also studied the spectrum of an interesting region of space, the Pleiades star cluster in Taurus, and in December of that fateful year detected the existence of dust within the cluster. In the same year that Slipher discovered the expanding universe, then, he also found that matter existed between the stars. He had, for the first time, detected what astronomers would come to call the interstellar medium. Before this time, astronomers had every reason to believe that the space between the stars was completely empty.

Nebula surrounding Merope

This is a more detailed picture of the reflection nebula NGC 1435, which surrounds Merope, one of the Pleiads. It was here that V. M. Slipher first detected the interstellar medium, working at Lowell Observatory in 1912.
Credit: Adam Block/Mount Lemmon SkyCenter/University of Arizona

Subsequent decades of research revealed that the interstellar medium consists of a wide range of matter and radiation, dust, and cosmic rays. It is the soup within a galaxy, if you will – all of the matter and energy that exists apart from star systems themselves. So, it turns out that all of our cosmic clouds, the nebulae, are a part of this interstellar medium. Astronomers talk about multiple phases of the interstellar medium, that is, that this material can be ionized (energized, like the gas inside most nebulae), atomic (free atoms), or molecular (gas tied up in molecules). Most of the interstellar medium consists of various forms of hydrogen, followed by helium, carbon, oxygen, nitrogen, and other elements.

Properties of the interstellar medium vary depending on place within a galaxy, temperatures, pressures, magnetic fields, and all manner of other factors. Although we can think of the interstellar medium as a free-ranging sea of particles throughout the galaxy, with "hot spots" represented by denser areas like nebulae, in general it is extremely thin. Cool, dense regions of the interstellar medium have typical densities of a million atoms per cubic centimeter. Hot, ionized regions like star-forming nebulae have densities of the order of 10,000 times less than this. (So only 100 atoms per cubic centimeter.)

Compare that to the typical density of a laboratory vacuum chamber, which contains something like ten billion atoms per cubic centimeter. So, although we can see lots of objects out in the interstellar medium, including nebulae, they exist as extremely low-density areas of matter. We're able to see them because they're large, and because they are often energized, or lit up.

We can envisage the interstellar medium as containing several components, as we understand it in the Milky Way. A significant volume within the disk consists of so-called Warm Ionized Medium, followed by the Warm Neutral Medium (containing atomic gas). Much smaller volumes are composed of the Cold Neutral Medium, followed by molecular clouds (containing gas in molecular form), and by HII regions, ionized areas of atomic hydrogen. Finally, a significant amount of material in the interstellar medium exists as coronal gas, also known as the Hot Ionized Medium, which lies far out in the galaxy's corona, surrounding the disk.

Our treasured nebulae all exist within the interstellar medium, and they show a variety of types and forms. In a primary sense, however, there are five broad types of nebulae. These classifications deserve a detailed examination, so that we can appreciate why nebulae exist and how they work throughout their lifetimes.

EMISSION NEBULAE

The most familiar category of nebulae are emission nebulae, also known less precisely as diffuse nebulae. The major type of emission nebulae are HII regions, also known as star-forming regions, and these are scattered throughout the sky, making up the majority of familiar nebulae that amateur astronomers observe in their telescopes.

The name HII region comes from the terminology astronomers use: HII stands for singly ionized hydrogen gas. Emission nebulae are made up of ionized gases and glow by fluorescence, in the same way a fluorescent bulb emits light. Their atoms are excited into glowing by the energetic stars within them and around them, and they are emitting light that we can see over very large distances. They are complex structures, consisting of a huge range of components, all jostled together in an environment that changes over time. A typical emission nebula of

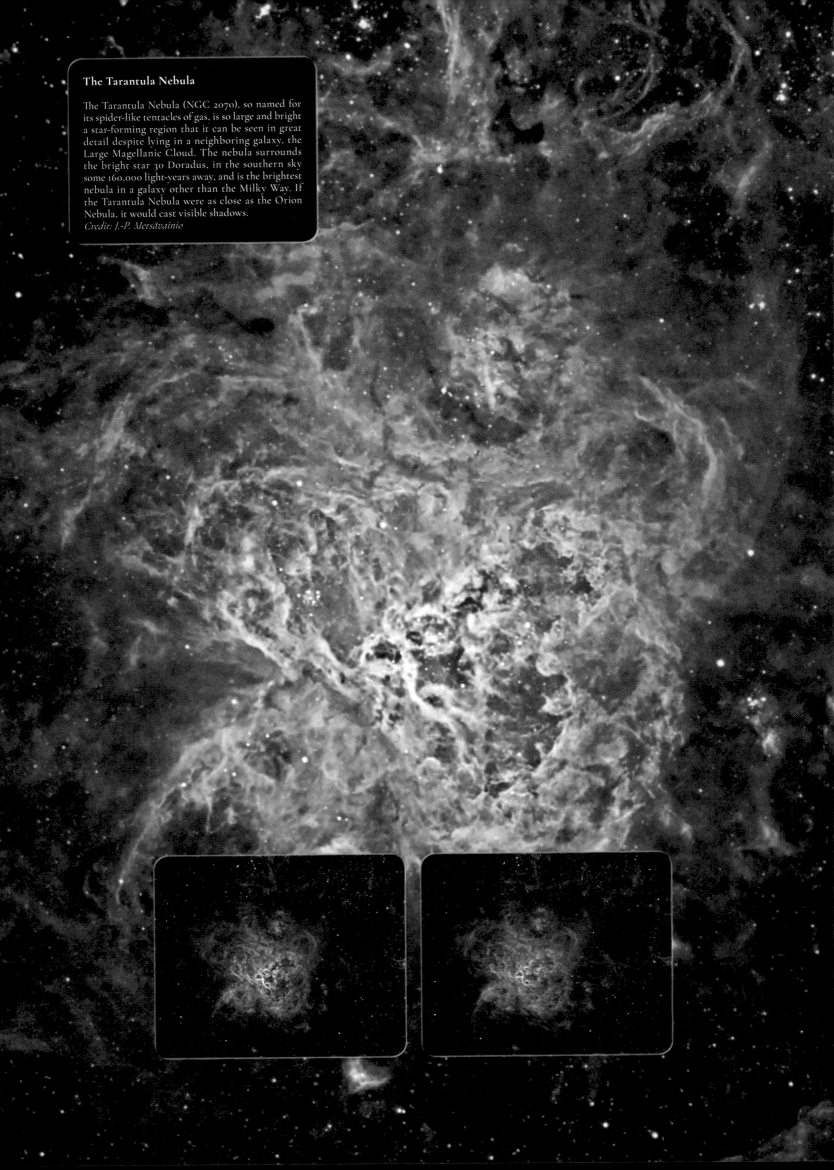

The Tarantula Nebula

The Tarantula Nebula (NGC 2070), so named for its spider-like tentacles of gas, is so large and bright a star-forming region that it can be seen in great detail despite lying in a neighboring galaxy, the Large Magellanic Cloud. The nebula surrounds the bright star 30 Doradus, in the southern sky some 160,000 light-years away, and is the brightest nebula in a galaxy other than the Milky Way. If the Tarantula Nebula were as close as the Orion Nebula, it would cast visible shadows.

Credit: J.-P. Metsävainio

the star-forming type contains lots of ionized gas, which we see as a misty light; a star cluster of young suns that are forming from the gas; a substantial amount of atomic gas that has not been ionized; and a large amount of associated dust.

Such star-forming emission nebulae may last for considerable millions of years as much of their gas gravitationally collapses into smaller spaces, forming thousands of stars, within a central cluster. Over such long periods, most all of the gas within an emission nebula may be converted into newborn stars, and much of the rest of it dispersed, as the game of interstellar recycling carries on. Moreover, tidal forces caused by the Milky Way's mass will eventually tear apart and disperse these young open clusters as they orbit the galactic center. So, over the course of millions of years, gas clouds form as the gas comes together, and they give birth to hot, young stars; the stars ionize the gas, causing it to glow, and eventually the whole starbirth process plays out, dispersing more new stars into the galaxy.

HII regions show a wide range of properties. They can exist as objects only a light-year across – about the size of our solar system – but most are far larger, up to several hundred light-years across. Nearly all the star-forming regions we can see well are in the Milky Way Galaxy, but these objects are sometimes bright and large enough to see in other galaxies. Notable examples of spectacular, distant star-forming regions include the Tarantula Nebula (NGC 2070), perhaps the largest known region of starbirth. It lies at a distance of about 160,000 light-years in the Large Magellanic Cloud, one of the Milky Way's satellite galaxies. Yet the Tarantula is still a bright telescopic object for southern hemisphere observers, spans at least 1,000 light-years, and if it were as close as the Orion Nebula (1,350 light-years), it would cast visible shadows on the ground at your next star party.

Another distant star-forming region of note is NGC 604, which exists in the Triangulum Galaxy (M33), and lies at a distance of 2.7 million light-years. Discovered by William Herschel in 1784, NGC 604 is bright enough to appear as a nebulous knot in backyard telescopes. It contains a cluster of some 200 stars, and is about the same size as the Tarantula Nebula.

Most familiar star-forming regions are closer to home, lodged relatively near to us in the Milky Way. The most familiar of all, to northern stargazers, is the Orion Nebula (M42). This misty naked-eye spot of haze opens up to a dazzling world of splendor for users of backyard telescopes. Known in antiquity as a puzzling, fuzzy "star," the Orion Nebula may have played a role in Mayan sky myths. Strangely, Galileo and several other early telescopists did not make mention of the nebula, but it was described in detail in 1610 by French astronomer Nicolas-Claude Fabri de Peiresc.

The Orion Nebula is a substantial star-forming region, stretching across about 25 light-years, and containing a cluster of newborn stars and protostars that will over the coming millions of years collapse into newly ignited suns. Lying at its relatively close distance of 1,350 light-years, the nebula provides a sterling laboratory for astronomers who want to observe the many stages of star formation. Not only do they see very young stars surrounded by abundant gas and dust, but also protoplanetary disks – solar systems in the making – brown dwarfs, and incredibly detailed evidence of turbulent gas that is helping to ionize the region, lit up by intensely hot young stars.

The deep southern sky also includes a spectacular HII region in the form of the celebrated, and physically much larger, Carina Nebula (NGC 3372).

Emission Nebula NGC 604

Many young stars can be seen in this image from the Hubble Space Telescope, along with what is left of the initial gas cloud. Some of the stars were so massive they have already evolved and exploded as supernovae. The brightest stars emit light so energetic that they create one of the largest clouds of ionized hydrogen gas known, comparable to the Tarantula Nebula in our Milky Way's close neighbor, the Large Magellanic Cloud.
Credit: NASA and the Hubble Heritage Team (AURA/STScI)

The Orion Nebula

A deep image of the Orion Nebula shows what is perhaps the favorite star-forming region in the sky, the most viewed by night-sky viewers, in all its glory. The region lies 1,350 light-years away and this star-forming cloud spans some 25 light-years. *Credit: Terry Hancock*

The Mouldy Strawberry Nebula

In the northernmost part of Orion lies a huge cloud of nebulosity called the Lambda Orionis Ring. A portion of this ring is Sharpless 2–265, which appears near the bottom of this image and is sometimes called the Mouldy Strawberry Nebula. It lies approximately 1,300 light-years away. *Credit: Alistair Symon*

Nebulae of the Southern Hemisphere

Deep within the southern hemisphere sky lie some of the greatest nebulae we know of, and the brightest is the Carina Nebula (NGC 3372), visible at lower right center in this image. The bright constellation Crux, the Southern Cross, stands at upper left. The reddish nebula left of center is IC 2944. A trio of bright star clusters forms a triangle around the Carina Nebula. They are NGC 3532 (above and left of the nebula), IC 2602 (below and left of the nebula), and NGC 3114 (below and right of the nebula). *Credit: Yuri Beletsky*

One of us (DE) often travels to a southern star party in Costa Rica, sponsored each year by *Astronomy* magazine. It is an astonishing thing to see the beautiful Orion Nebula perched high in the sky, and then look over – in the same sky at the same moment – to see the Carina Nebula, appearing far larger and brighter than the familiar friend in Orion.

The Carina Nebula lies at the greater distance of about 8,500 light-years – some six times farther away than the Orion Nebula – yet it is far brighter and spans about 230 light-years across. Physically, it's almost ten times larger than the Orion Nebula. So even at its greater distance, it appears commandingly larger and brighter in our sky. It is the most impressive emission nebula we can see from planet Earth.

The Carina Nebula appears like a brilliant, blocky glow that is sliced into several parts by broad swaths of dark nebulosity floating in front of the emission gas. Visible in antiquity to the naked eye, the nebula spans about four times the width of the Full Moon and was first written about in detail by French astronomer Nicolas-Louis de Lacaille in 1752. It is one of the most glorious sights in the deep southern sky, along with the Magellanic Clouds, two satellite galaxies of the Milky Way.

The Carina Nebula

The brightest emission nebula in our sky is the Carina Nebula (NGC 3372), a vast complex of bright and dark features in the deep southern sky. This enormous star-forming region lies about 8,500 light-years away and contains several important features, including the enigmatic star Eta Carinae. This odd variable star, embedded within the nebula, was a brilliant naked-eye star in the mid-19th century before fading to its present brightness. The nebula also contains several star clusters.
Credit: Tony Hallas

PLANETARY NEBULAE

Another major type of nebulae are called planetary nebulae, a specialized class of emission nebulae. Although planetary nebulae glow like HII regions, they are very different types of nebulae. Consequently, astronomy enthusiasts think of them as distinctly different types of objects, as they represent different ends of stellar evolution. Whereas most emission nebulae, HII regions, are stellar birthplaces giving rise to new generations of suns, planetary nebulae are the endpoints, the cocoons of dying stars that have the same relative mass as the Sun.

When solar-mass stars run out of hydrogen fuel to burn, they enter the later stages of life, and eventually become so-called red giant stars, swelling up in size. They then belch off layers of gas that become visible as a planetary nebula, glowing by fluorescence for approximately 50,000 years, until the dying central stellar core becomes an ultra-dense white dwarf.

Planetary nebulae have nothing to do with planets, but appeared as glowing disks, similar in appearance to planets, in early telescopic observations. So, the great German-English observer William Herschel named this class planetary nebulae in the 1770s.

Some 3,000 planetary nebulae exist in the Milky Way, although most are small and relatively faint, and are somewhat difficult to see over large distances. However, several hundred are notable examples that are bright and/or close enough to capture magnificent detail within. One of the most prominent in the northern sky is the Ring Nebula (M57) in Lyra, which appears in small telescopes as a perfect "smoke ring" of glowing, grayish-green light. The central star responsible for creating the nebula is relatively faint – bright enough to photograph easily but a

William Herschel

On 13 March 1781 William Herschel discovered a new object in the constellation of Taurus, which was eventually confirmed to be a new planet – Uranus. This was his most famous discovery and led to him being given the position of Court Astronomer by King George III. Over the next 20 years or so he discovered and cataloged over 2,400 "nebulae". At that time nebula was a term for any heavenly object that was fuzzy and not a star, and embraced galaxies as well as every category of nebulae. *Credit: National Portrait Gallery/Public Domain.*

The Ring Nebula

One of the most recognizable planetary nebulae is the Ring Nebula (M57) in Lyra, a favorite target for observers armed with small telescopes. This deep exposure of the area shows the characteristic smoke-ring shape of the nebula that gave rise to its name, but also a much fainter ring of gas surrounding it. The Ring Nebula lies some 2,600 light-years away, and the galaxy IC 1296, above and left of the nebula, lies 220 million light-years away – some 88,000 times farther away.
Credit: Tony Hallas

challenge to see visually in a telescope.

Discovered by the French comet hunter Charles Messier in 1779, the Ring Nebula lies about 2,600 light-years away and spans about 1.3 light-years across – roughly the size of our solar system. The nebula appears to us as a prolate spheroid, like a rugby ball, but it is in fact bipolar, and shows faint outer rings of nebulosity far larger than the bright ring shape we see in a telescope.

Not far away in our sky from the Ring is another prominent planetary nebula, the Dumbbell Nebula (M27) in the small constellation Vulpecula. As its name suggests, this nebula has a prominent dumbbell shape, easily visible even in small telescopes. Lying at a distance of 1,350 light-years, this bright nebula spans nearly 1.5 light-years across, and is also shaped like a prolate spheroid. We view the dumbbell along the plane of its equator; that is, we are seeing it "end-on." Studies of the expansion of the Dumbbell Nebula suggest it is about 10,000 years old. It was the first planetary nebula to be found, before the coining of the term planetary nebula, by Charles Messier, in 1764.

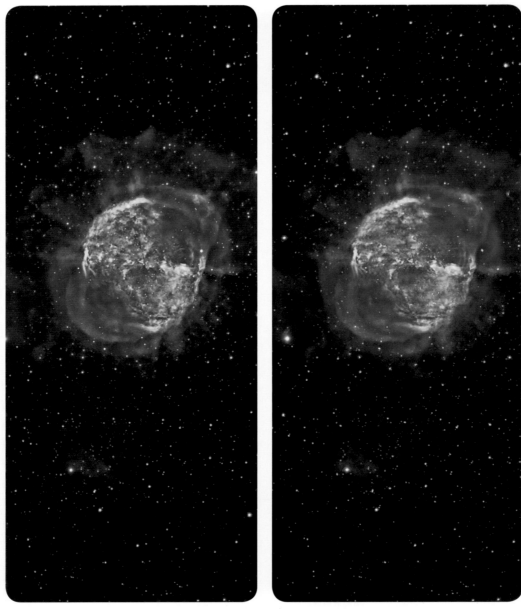

The Dumbbell Nebula in 3-D

Viewing the distinctive Dumbbell Nebula (M27) in stereo, the dumbbell shape appears to transform into the physical reality of the object, a bloated sphere. Such planetary nebulae form from low-velocity gas belched away from the parent star at first, and then higher-velocity ejections. The collision between the gas clouds helps to light them up, giving us such a majestic view.

The Dumbbell Nebula

Planetary nebulae, so named by William Herschel because in telescopes their glowing disks resemble planets, are the destiny of stars similar to the Sun. At the ends of their lives, Sun-like stars run out of fuel and puff their remaining layers off into space, creating these glowing cocoons of light. The Dumbbell Nebula (M27) in Vulpecula might approximate what the Sun will look like five or six billion years from now.

Credit: R. Jay GaBany

Planetary Nebula Abell 36

Abell 36 is a bluish planetary nebula lying about 780 light-years distant in the constellation Virgo. Its bright central star is visible in backyard telescopes, and illuminates the softly glowing cloud that surrounds it. *Credit: Adam Block/Mount Lemmon SkyCenter/ University of Arizona*

Abell 71 and Sharpless 115

This photo shows about one square degree of sky from the constellation Cygnus. In it there is both birth and death. The bluish central area is a stellar nursery and the blue dot at lower left is the planetary nebula Abell 71. It has consumed all its hydrogen resources and blown off its outer layers.
 Credit: J.-P. Metsävainio

REFLECTION NEBULAE

Emission nebulae glow like fluorescent light bulbs. The third major class of nebula does not glow at all. Rather, a reflection nebula is a cloud of dust and often associated gas that we can see as a glow in our sky only because it reflects starlight from a nearby sun toward us. Thus, reflection nebulae are relatively rare compared with bright nebulae, because they must be in close proximity to bright stars and also aligned in the right way to enable us to see them. So, while there are thousands of bright nebulae in the sky, only hundreds of reflection nebulae exist to telescopic observers, and the choice ones can be counted in the dozens.

The dust grains that make up reflection nebulae are tiny, and we see

The Iris Nebula

The Iris Nebula (NGC 7023) in Cepheus is a prominent example of a reflection nebula, its dusty grains redirecting starlight from the 7th-magnitude star SAO 19158, embedded within it. Lying some 1,300 light-years away, the nebula spans some six light-years, about three times the diameter of our solar system. *Credit: Tony Hallas*

these nebulae only because they are enormous. Moreover, they need to be in association with stars at the right distances, such that associated gas is not ionized, but the dust merely reflects. The spectra of reflection nebulae, therefore, are in lockstep with the kinds of stars that illuminate them. The tiny particles scattering the starlight are believed to be carbon-rich compounds, as well as other simple elements such as iron and nickel. Because iron and nickel dust is typically aligned with the magnetic field of the galaxy, the scattered light from a reflection nebula is often polarized.

Examples of reflection nebulae lie scattered across the sky. One of the best known is M78, a bluish cloud in Orion, not far north of the Orion Nebula itself. Lying at a similar distance to the Orion Nebula, 1,350 light-years, M78 is a dusty portion of the Orion Molecular Cloud that sweeps over much of that constellation, and is illuminated by bright blue stars of spectral type B.

M78 in Orion

M78 is a bright reflection nebula in Orion, lying north of Orion's Belt and some 1,350 light-years away. Two relatively bright blue-white stars are embedded within the nebula, and their light is reflected toward us through M78's dusty cloud. *Credit: Tony Hallas*

NGC 6726

Bluish reflection nebulae flood the field of this image of the southern constellation Corona Australis. The bluish areas of dust surrounding the bright stars at top are NGC 6726 and NGC 6727; reflection nebulosity surrounding the bright blue double star near the bottom is IC 4812. The butterfly-shaped, orange-colored emission nebula left of center is NGC 6729, and it surrounds the variable star R Coronae Australis. As the star varies in brightness, the nebula also appears to grow and shrink in size and brightness.
Credit: Gerald Rhemann

DARK NEBULAE

Even murkier than reflection nebulae is the next category of cosmic cloud. A dark nebula is an interstellar cloud with sufficient density to block the light from stars and other luminous objects behind it. Sometimes also called absorption nebulae because they absorb light from beyond, dark nebulae are visible, interlaced all throughout the Milky Way's glowing band, on a very dark, moonless night from a perfect observing site.

Composed of tiny dust grains about the size of the particles in cigarette smoke, dark nebulae are visible because they are placed in front of luminous objects; they are very dense, they are very expansive, or a combination thereof. Originally thought to be "holes in the sky" by early observers, the world of dark nebulae was transformed by American astronomer E. E. Barnard, who studied them and photographed them extensively in the late 19th and early 20th centuries, culminating in his famous work *A Photographic Atlas of Selected Regions of the Milky Way*, published in 1927.

Gyulbudaghian's Nebula

This spectacular and rich field of stars and dark nebulosity in Cepheus hides a secret: Gyulbudaghian's Nebula, a strange reflection nebula that varies in size and brightness. The nebula appears as a tiny triangular smudge within the dark nebula directly above the bright star at bottom center. Armenian astronomer Armen Gyulbudaghian discovered this nebula in 1977, and it varies as the embedded star brightens and fades. *Credit: Thomas V. Davis*

Among the best-known dark nebulae is B33 (many dark nebulae carry Barnard numbers, after the pioneering astronomer), the Horsehead Nebula in Orion. Also a part of the Orion Molecular Cloud complex, the equine-shaped nebula is small and difficult to see in backyard telescopes, but can be photographed far more easily, set off against a faint strip of glowing emission nebulosity. The Horsehead lies about 1,400 light-years away and was first described in detail by Scottish-American astronomer Williamina Fleming, who photographed the nebula at Harvard College Observatory in 1888.

Williamina Fleming

Williamina Paton Stevens Fleming (1857–1911), c.1890s. *Credit: Curator of Astronomical Photographs at Harvard College Observatory/Public Domain*

The Horsehead Nebula

One of the sky's best-known dark nebulae is the Horsehead Nebula (B33), so named for its distinctive, equine-like shape. Dark nebulae are composed of dust grains and they are visible only when backlit by brighter objects. The Horsehead Nebula lies in the constellation Orion, at a distance of about 1,400 light-years, and is a difficult object to spot visually in backyard telescopes.

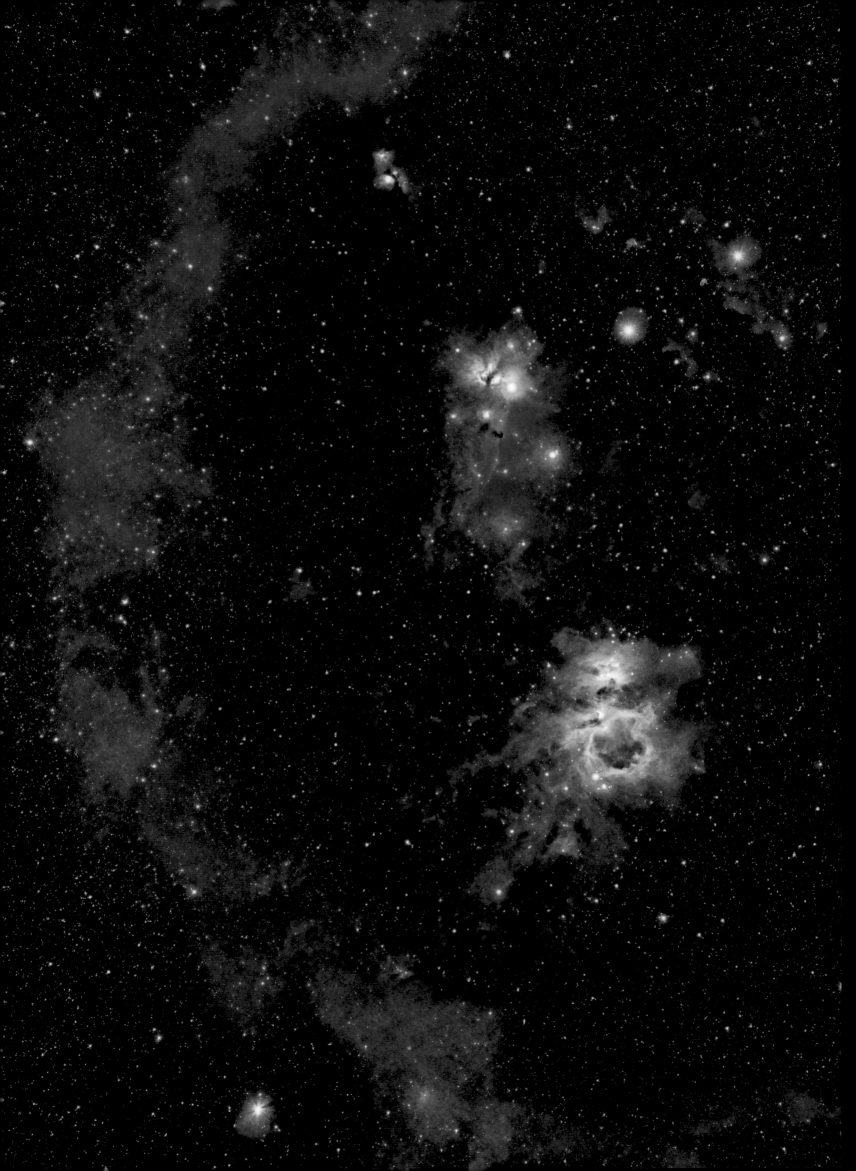

Barnard's Loop

A sweeping view of the lower portion of Orion shows multiple nebulae, including Barnard's Loop, the reddish, C-shaped arc along the left edge. It resulted from a supernova explosion some two million years ago. Slightly below and left of center is the Orion Nebula. On the left side of Orion's Belt, left of center, lies the Horsehead Nebula, a tiny black dust cloud cutting into the strip of reddish emission nebulosity.
Credit: Terry Hancock

The Crab Nebula

The Crab Nebula (M1) lies at a distance of some 6,500 light-years and spans about 11 light-years. its gas is now expanding at a rate of about 1,500 kilometers per second.
Credit: J.-P. Metsävainio

SUPERNOVA REMNANTS

The final broad class of our cosmic clouds represents an unusual endgame for massive stars. For stars with about eight to several hundred times the mass of our Sun, the star's fate will not be a planetary nebula, but rather a supernova. This event, one of the most violent and energetic known in the universe, results from one of several processes that takes place when a massive star runs out of nuclear fuel. The star explodes, sending a cloud of debris expanding rapidly outward into the surrounding galaxy. The result is a supernova remnant, and many of these clouds of catastrophic explosions are visible around the galaxy.

The most famous and historically important such object is the Crab Nebula (M1) in Taurus, visible in small telescopes as an oval smear of glowing nebulosity. Named for the numerous "tentacles" of gas visible in long-exposure photographs, the Crab Nebula is the result of a star that exploded in the year 1054. That is known with precision because Chinese observers left detailed accounts of observing a bright, daytime star in that position of the sky.

Another celebrated supernova remnant is the Veil Nebula (NGC 6960, NGC 6979, and NGC 6992–5) in Cygnus, a huge, spherical shell in which several edges of the expanding cloud are lit up nicely as they are interacting with the interstellar medium (*see also* chapter 6). The star that produced the Veil Nebula exploded about 8,000 years ago and the whole cloud lies at a distance of about 1,470 light-years, giving it a diameter of some 100 light-years. Such clouds of glowing gas remind us of the interstellar recycling program that is always going on as some stars are born and others die. The population of the galaxy is ever changing, albeit far more slowly, than the population of humans and our fellow creatures on Earth.

Indeed, in the galaxy, distances are so vast and the cosmic timescale so huge, that we see in effect an enormous and indescribably complex movie, but we only see it one frame at a time.

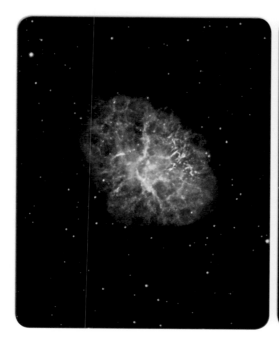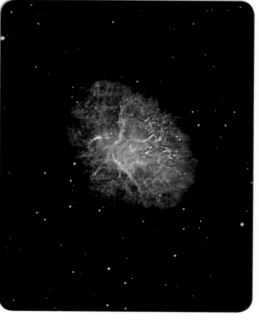

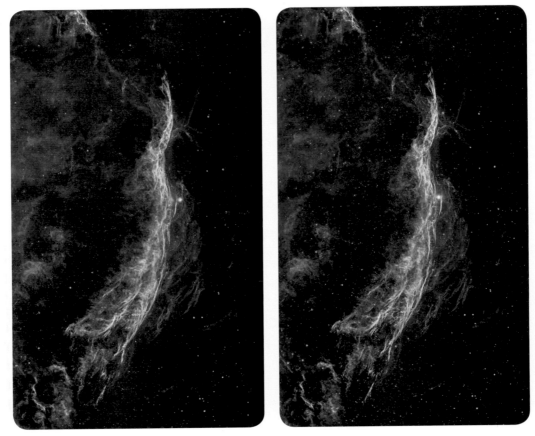

The Witch's Broom Nebula

This western segment of the Veil Nebula is cataloged as NGC 6960. It spans
35 light-years, and the bright star 52 Cygni can be seen with the naked eye.

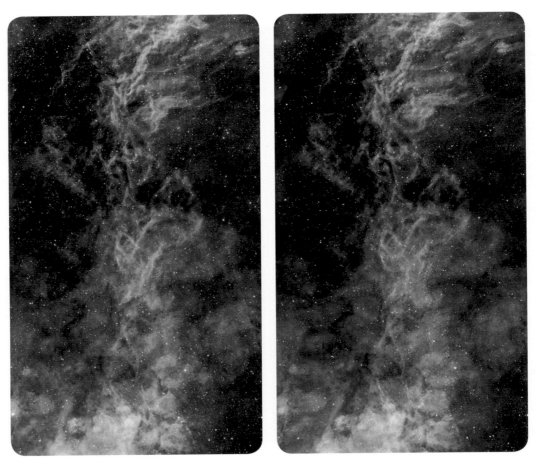

Central Cygnus in 3-D

This mosaic of filaments in central Cygnus spans about two degrees of sky. The image is mapped in
colors emitted by ionized elements – red is sulfur, green is hydrogen, and blue is oxygen.

Central Cygnus

The center of the constellation Cygnus gleams in this long-exposure photograph, showing the northern constellation riddled with reddish emission nebulosity. The bright blue-white star Deneb lies at top center; just to its left are the North America and Pelican nebulae. Between Deneb and the center of the frame is the huge complex of nebulosity cataloged as IC 1318. The Veil Nebula, one of the sky's best supernova remnants, lies at lower left.
Credit: Alistair Symon

Orion Head to Toe

A magnificently detailed image of the constellation Orion reveals a multitude of nebulae, including Barnard's Loop arcing through the left side; the Lambda Orionis Nebula at top; the Orion Nebula below center; the Horsehead Nebula just below and left of center; and the Witch Head Nebula at bottom right.
Credit: Rogelio Bernal Andreo

LIFE CYCLES OF THE STARS

4

I n one sense stars are born on Hollywood Boulevard, but in the universe at large, they are born in giant molecular clouds. As we've seen, these large clouds of molecular gas typically contain 1,000 to 10 million solar masses and stretch over diameters of dozens to hundreds of light-years. As gravity squeezes their matter down into smaller spaces, denser portions of the cloud emerge, in the form of clumps, bubbles, or sheets, with less dense areas between them.

We can see evidence of these molecular clouds, in some cases relatively nearby. The Orion Molecular Cloud covers much of that bright winter constellation, enshrouding its form with numerous nebulae. That cloud consists of two main parts and is an ongoing scene of active star formation, with the newborn suns in the range of 12 million years old and younger. The cloud stretches several hundred light-years across and lies some 1,000 to 1,400 light-years away, with several of its most important pieces, including the Orion Nebula, lying about 1,350 light-years distant.

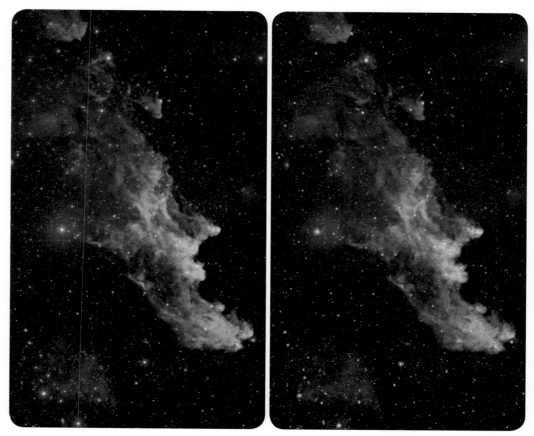

The Witch Head Nebula

IC 2118 in Eridanus is extremely faint to the eye alone but easily photographed with long exposures. It lies about 900 light-years away and reflects starlight from Rigel, the brightest star in Orion.

LDN 1622 in 3-D

In stereo this dark nebula in Orion appears like a dusty, cosmic horse,
and also reveals a broad background draped in nebulosity.

Dark Nebula in Orion
The creature-like shape of blackness that
emerges from the center of this image
represents LDN 1622, a dark nebula in Orion,
located not far from a portion of Barnard's
Loop. The emission nebulosity that helps
to outline LDN 1622 is van den Bergh 62.
Credit: Daniel B. Phillips

Barnard's Loop

Part of Barnard's Loop (Sh 2-276), the prominent supernova remnant, arcs across this wide shot in the Orion Molecular Cloud. The reflection nebula M78 in Orion lies at bottom left, and top right is LDN 1622 (also seen opposite), sometimes called the Bogeyman Nebula.
Credit: Alistair Symon

Bok Globules

The great naked-eye emission nebula IC 1396 in Cepheus lies adjacent to the reddish star Mu Cephei (right edge of the nebula), known as Herschel's Garnet Star. This splendid object contains a multitude of dark nebulae, objects collapsing down into newborn stars, called Bok globules. One of the most famous dark nebulae in the area is known as the Elephant's Trunk (reaching from the bottom of the nebula toward its center). *Credit: J.-P. Metsävainio*

The Elephant's Trunk Nebula in 3-D

The Elephant's Trunk Nebula winds through the emission nebula and young star cluster complex IC 1396, in the constellation Cepheus. Also known as vdB 142, the cosmic elephant's trunk is over 20 light-years long.

Closer yet is the Taurus Molecular Cloud, a vast area covering portions of the constellations Taurus and Auriga in our sky, and lying a mere 430 light-years away. As such, it is the nearest large star-forming region in the galaxy. These star-forming regions lie within our local spiral arm of the Milky Way, called the Orion Arm, and in fact a bright group of stars is arrayed in a ring that helps to define this arm. Called Gould's Belt, it was discovered in 1879 by American astronomer Benjamin A. Gould. Containing many hot, young, type O and B stars, this belt is between 30 and 50 million years old, and is about 3,000 light-years across.

Looking at our nebulae, star-forming regions, scattered across the sky, we can see widespread evidence of small molecular clouds, too. Many such areas lie within star-forming regions and have masses less than a few hundred times that of the Sun. These appear as little dark areas in and around HII regions, and they are called Bok globules, after Dutch-American astronomer Bart Bok, who studied them extensively in the 1940s. These little clouds of dust and gas are in the latter stages of collapsing toward making new stars. Often, they contain from two to several dozen solar masses of material in a region a light-year or so across, about the size of our solar system. Frequently, they create double or multiple star systems.

Starbirth

As molecular clouds collapse by gravity, they separate into smaller pieces, and in each of the fragments, pressures and temperatures rise as gravitational potential energy is released. Infalling motions create a rotational effect, and when enough matter falls inward, a critical mass eventually is reached, and a protostar is born. This rotating sphere of hot gas is a star in the making. It's still very young and is still taking on more mass from the surrounding material in the molecular cloud. This, however, is the earliest phase of what astronomers call stellar evolution, the understanding of how stars form and change over time.

After some period of additional time, hundreds of thousands of years, the protostar depletes the surrounding gas and an opaque core forms inside the star, which consists of course mostly of hydrogen gas. For a star of the Sun's mass or less, this process takes place on the order of about half a million years. When the infalling gas is gone, the object becomes a pre-main sequence star. This is a stage in stellar evolution when a newborn star has not yet reached the so-called main sequence stage. This term refers to the normal stage of a normal star, as defined by a plot of a star's color and magnitude, called the Hertzsprung-Russell (HR) diagram or Color-Magnitude Diagram (CMD). This plot shows a distinctive band of stars running in a line, representing the normal hydrogen-burning phase of most stars. It is therefore called the main sequence.

As our pre-main sequence star prepares itself for an ordinary stellar life, its energy generally blows away any surrounding gas and dust in its immediate vicinity. It then becomes visible to us over large distances as a young star in the making. It has gathered together nearly all of its mass, in hydrogen, mostly, but has not yet "turned on" fully by beginning nuclear fusion. As it continues to collapse, its temperature still rising, the star finally reaches a critical point and finally begins fusing hydrogen, landing itself on the main sequence, and becoming a fully-fledged star.

It may have just gone through one of a few stages, such as being a so-called T Tauri star, if it had fewer than two solar masses. Or it may have been classed as a Herbig Ae/Be star, if it had between two and eight solar masses. Stars more massive than this do not undergo a pre-main sequence phase at all, as they collapse very rapidly and initiate hydrogen fusion in their cores as they become visible.

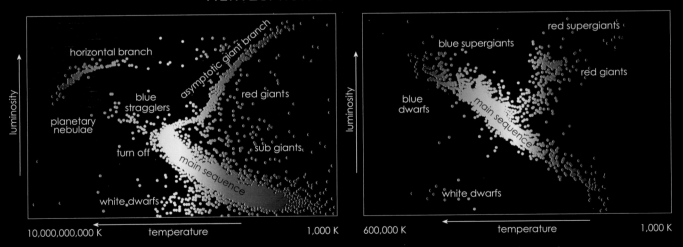

The Hertzsprung–Russell diagram, abbreviated as H–R diagram, is a scatter plot of stars showing the relationship between the stars' luminosity (brightness) and temperature (color). The left graph shows the pattern for older stars and the right for younger stars. As stars evolve they move up the main sequence from lower right to upper left, and where they end up depends upon their mass.

GENESIS OF A SOLAR SYSTEM

As these stars contract and begin hydrogen fusion, they are invariably rotating, as a byproduct of the physics and energy of the systems. Material that has not been drawn completely into the star inevitably remains, theorists believe, and ends up commonly as a rotating cloud of debris called a circumstellar disk. In fact, we see evidence of circumstellar disks, the clouds that form planets and small bodies like asteroids and comets, around many young stars nearby in the galaxy.

These pancake-shaped clouds are visible in many sites of star formation, such as in and around the Orion Nebula, which contains numerous protostars. This of course only makes complete sense and is consistent with the observations of many nearby stars in our galactic neighborhood, where astronomers have discovered several thousand extrasolar planets, and the count is rising every week.

Our own solar system holds a multitude of components – not only the Sun and our planets, but dwarf planets too, and other major components. The main belt of asteroids between Mars and Jupiter comprises probably two million objects. The Edgeworth-Kuiper Belt, icy bodies beyond the orbit of Neptune, contains millions of objects far out in the solar system. The Scattered Disk contains a population of icy objects in weird orbits, scattered hither and yon by the gravity of the ice giant planets. And the Oort Cloud, on the physical edge of our solar system, contains perhaps two trillion cometary nuclei, some of which occasionally fall in toward the Sun.

Astronomers have every reason to believe that all star systems would have similar components, even if in different configurations.

Once a star begins hydrogen fusion, it generates immense amounts of light and heat as a natural nuclear reactor, and affects everything within its system and beyond in profound ways. First and foremost, for us, it makes life possible. The Ancient Egyptians were onto something: the Sun is the ultimate source of all life on Earth. Not all stars in the universe are main sequence stars, but the majority are, including the Sun.

Where astronomers place stars in terms of their evolution on the main sequence depends on several factors, including mass, chemical composition, and age. Such stars burn hydrogen, converting it into helium and then heavier elements, in a predictable and progressive way, because they are in balance. This state of hydrostatic equilibrium, as it's called, comes about because the inward pressure of gravity balances against the outward pressure of heat and energy escaping from the hot core.

Astronomers divide the main sequence into two parts, because stars of about 1.5 or fewer solar masses fuse hydrogen in multiple stages to form helium. Astronomers call this process the proton-proton chain. More massive stars employ a fusion process that uses atoms of carbon, nitrogen, and oxygen as intermediaries, and so this CNO Cycle, as it's known, produces helium. Less massive stars have more pure radiative effects going on in their cores, while more massive stars have convective centers that stir up the helium and help to propel the fusion process forward.

More massive stars have shorter lifetimes than less massive stars. With stars, it truly is "live fast and die young" for the massive, furiously burning suns. The least massive stars radiate slowly and can last almost forever.

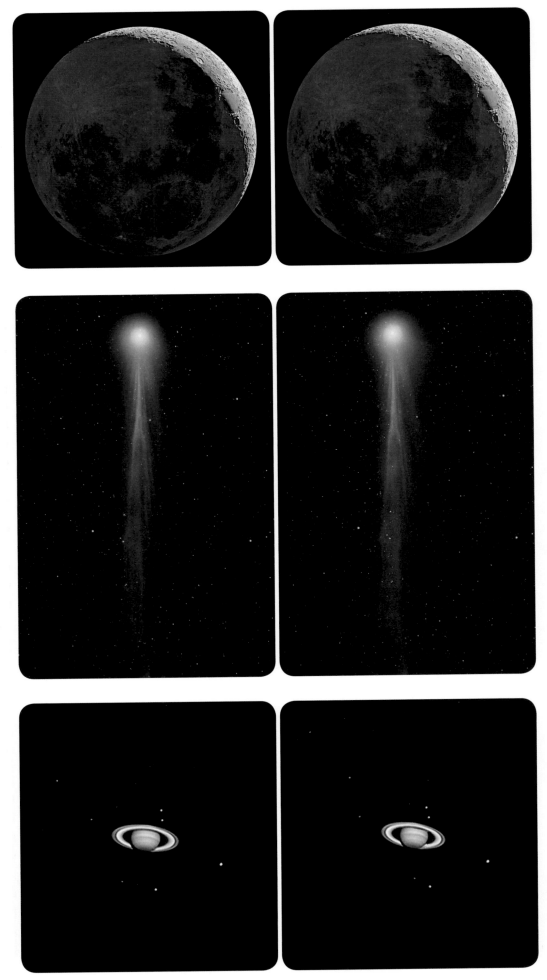

Snapshots from the Solar System in 3-D

Top: the Crescent Moon; *Middle*: Comet Lovejoy; *Bottom*: the Moons of Saturn.

CLASSIFYING THE STARS

A star's mass is important in other ways, too. Stars are classified by their masses and their temperatures. Their masses are measured relative to the mass of our Sun, which defines one solar mass, and their temperatures are expressed in kelvin, where 0 degrees kelvin (K) is equal to minus 273 degrees Celsius (C). More than a century ago astronomers at Harvard College Observatory worked furiously to classify millions of stars. At Harvard, astronomer Annie Jump Cannon formalized the basics of the current system. By the mid-20th century, astronomers W. W. Morgan and Philip C. Keenan, at Yerkes Observatory, refined the understanding of stellar spectra. Stars are divided by their colors, including classes O, B, A, F, G, K, and M. For decades, astronomers remembered the composition by the mnemonic "Oh, Be A Fine Girl/Guy, Kiss Me."

The hottest stars are O stars, blue suns with temperatures at or greater than 30,000 kelvins, and masses

Annie Jump Cannon

Annie Jump Cannon (1863–1941) was an American astronomer at the Harvard College Observatory whose cataloging work was instrumental in the development of contemporary stellar classification. *Credit: New York World-Telegram and the Sun Newspaper/Public Domain*

on the main sequence of at least 16 solar masses. The next class, B stars, are blue-white in color, have temperatures of 10,000 to 30,000 kelvins, and main sequence masses of 2 to 16 solar masses. Then come A stars, white stars, with temperatures of 7,500 to 10,000 kelvins, and main sequence masses of 1.4 to 2.1 solar masses. Next are the F stars, yellow-white in color, with temperatures of 6,000 to 7,500 kelvins, and main sequence masses of 1 to 1.4 solar masses. And then come the G stars, which are yellow, have temperatures of 5,200 to 6,000 kelvins, and main sequence masses of 0.8 to 1 solar mass. Our Sun is a G-type star. Next come the K stars, which are light orange, have temperatures of 3,750 to 5,200 kelvins, and main sequence masses of 0.5 to 0.8 solar masses. And lastly come the M stars, which are orange-red, have temperatures of 2,400 to 3,700 kelvins, and main sequence masses of 0.1 to 0.5 solar masses.

Moreover, classifying stars by their spectra, or colors, can also reveal something about their physical state. At Yerkes, Morgan and Keenan, along with fellow astronomer Edith Kellman, developed a system of analyzing spectra and understanding not only temperature (color) and luminosity of a star, but also surface gravity, which is related to luminosity. Denser stars with higher surface gravity show broadening of their spectral lines, and by the early 1950s, this system came to be known as the Morgan-Keenan classification scheme. It defines ten types of stars, as follows, from the brightest, largest, and most fiercely burning stars, to the least:

LUMINOSITY CLASS	DESCRIPTION	EXAMPLE STARS
0 or Ia⁺	hypergiants	Eta Carinae
Ia	luminous supergiants	Rigel
Iab	intermediate-size supergiants	Sadr
Ib	less luminous supergiants	Zeta Persei
II	bright giants	Nihal
III	normal giants	Arcturus
IV	subgiants	Gamma Cassiopeiae
V	main-sequence stars	Our Sun
VI	subdwarfs	Kapteyn's Star
VII	white dwarfs	Sirius B

Classifying Stars by Luminosity

SPECTRA

Sir Isaac Newton was the first to realize that light could be split up into a spectrum of wavelengths from red to violet by passing a sunbeam through a prism. In 1801 English scientist William Wollaston showed the Sun's rainbow spectrum contained a number of dark lines, and later the German physicist Joseph von Fraunhofer discovered that these lines never changed their position in the spectrum, nor their intensity. But he did not know what they were. In 1858 Gustav Kirchhoff and Robert Bunsen solved the problem – each dark line in the Sun's spectrum corresponds to a bright line emitted by a particular element if it is burned in a laboratory (where it is known as an emission spectrum).

The lines we see in the solar spectrum appear dark because these bright lines are absorbed by that same element in gaseous form in the Sun's atmosphere, because the gas there is cooler and its electrons absorb energy at that characteristic frequency.

Named Fraunhofer lines, observing them reveals which elements are present in the Sun; this technique is known as spectroscopy and is very widely used in modern astronomy.

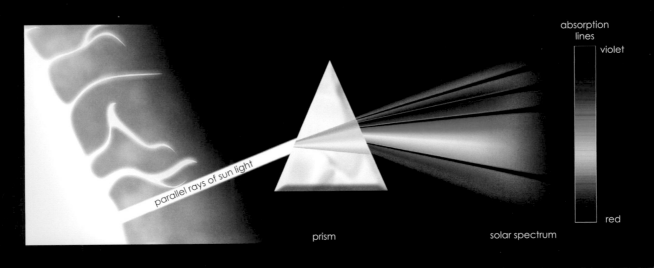

parallel rays of sun light

prism

solar spectrum

absorption lines

violet

red

Spectra are really the key to understanding stars, and all manner of objects in the universe. In addition to the previous broad categorizations, astronomers can understand a multitude of properties of stars based on numerous features and oddities observable in a star's spectrum.

Understanding the fundamental differences in the types of stars in the universe is important, as astronomers believe the story of how stars form is slightly different for stars of high mass and low mass. Hypotheses about star formation have been worked on for decades, and married with a growing number of observations. Astronomers believe low-mass stars form by the gravitational collapse of dense regions within molecular clouds. This collapse, as we've seen, leads to rotation, and the formation of a protoplanetary disk. Material continues to flow into the protostar during this process, and also some ends up as planets and smaller bodies once the star ignites.

For stars of more than about eight solar masses, however, the formation process is a little less well understood. Such massive stars are huge emitters of radiation pressure, an outward flow of radiation that pushes on surrounding material. For many years, astronomers wondered about radiation pressure inhibiting the formation of massive protostars. But theoreticians have worked extensively on understanding that perhaps channels form in the outflow of material from massive protostars that allow the radiation pressure to escape without preventing the accretion of material inward, onto the forming star, adding to its mass. So, astrophysicists believe that massive stars may form in a similar way to less massive stars, with this important difference, that allows radiation pressure to escape.

This mechanism is not well established, however, and some astronomers wonder whether massive stars can form when less massive stars "seed" a particular area of the molecular cloud, drawing in lots of matter. Or, alternatively, when two or more less massive stars form in close proximity and then combine into a larger, single entity.

Once stars form, they are off and running, converting hydrogen into helium, happily "burning" away as nuclear fusion reactors. As mentioned, more massive stars live fast and die young, converting their fuel at a ferocious rate. Lower mass stars last for an incredibly long time. For example, an O-type star with 60 solar masses has a lifetime of about three million years. An O-type star of 30 solar masses lasts for some 11 million years. A B-type star with ten solar masses lasts for about 32 million years. An A-type star with three solar masses has a lifetime of about 370 million years. An F-type star with 1.5 solar masses lasts for some three billion years. A G-type star like the Sun has a lifetime of about ten billion years. And an M-type star, like the ubiquitous red dwarfs, has a mass about one-tenth of the Sun's but can last for trillions of years.

STARDEATH

When a star exhausts its hydrogen supply, it migrates off the main sequence. Without the outward radiation pressure generated during fusion burning of hydrogen, the star's core collapses. For low-mass stars, no one knows exactly what happens when this process occurs, because the universe is 13.8 billion years old and fusion burning in low-mass stars can last for several trillion years. But astronomers understand the theory of what will happen. Red dwarfs, the most common stars in the universe, can slowly

The Giant Squid

This dim planetary nebula, known as OU4, lies in the constellation Cepheus, and really does resemble a giant squid. Vertically, seen from Earth against the background sky, it spans the same distance as three of our Moons placed side-by-side.
Credit: J.-P. Metsävainio

radiate like embers in a fireplace, eventually converting almost the entire star into helium. Eventually these stars will degenerate into white dwarfs.

Slightly more massive stars will expand into red giant stars near the ends of their lives, but they lack the mass to begin helium burning. A star of about half the mass of the Sun will become a red giant for a relatively brief period, and then transform into a white dwarf.

The story of aging stars we must often think of is the one that relates to our own Sun. Mid-sized stars of half a solar mass to ten solar masses will become red giant stars when they move off the main sequence. A star like this will exhaust hydrogen in its core and begin to fuse hydrogen in a shell outside its core, thereby leaving the main sequence. This is called the subgiant phase. It then enters the red-giant-branch phase, in which the helium core continues to grow and the star increases in luminosity. Later, the star enters the horizontal branch when the star experiences a helium flash and begins fusing helium. After the star has exhausted the helium in its core, hydrogen and helium fusion continues in shells around a core of carbon and oxygen. This is the so-called asymptotic-giant-branch phase.

Finally, these mid-range stars reach the end of the asymptotic giant branch on the H-R diagram, and they exhaust their fuel, unable to burn any further shells. They are not massive enough for another stage of fusion, carbon burning, and so they contract again. This process creates several "belches" of gas flowing away from the contracting star, and it is the mechanism that produces a planetary nebula.

The process of massive star evolution is very different, however. When a massive star evolves toward the end of its life, the star is massive enough that when the hydrogen shell burning occurs, helium ignition takes place, and these stars expand and cool, but they do not brighten substantially like lower-mass stars. Their cores cannot support themselves any longer and eventually they will become either neutron stars or black holes.

The most massive stars, those greater than 40 solar masses, lose mass rapidly from fierce stellar winds and thus lose their own envelopes before they could expand to become red supergiants. So, they remain extremely hot and bluish in color throughout their old age. The core of a massive star becomes hotter as the star evolves, and begins burning successively heavier elements. Temperatures become high enough to fuse carbon and heavier elements, and fusion of neon, sodium, and magnesium occurs. Depending on various factors, a star of about 8 to 12 solar masses can become highly unstable during this phase and the result can be an electron-capture supernova, a violent stellar explosion.

More massive stars yet experience oxygen burning, followed by fusion of neon and silicon. The core of such a star is now dominated by iron-peak elements (iron and elements with atomic numbers close to iron), and lighter elements surround the core, still undergoing fusion. The iron core grows until it reaches the so-called Chandrasekhar mass, which ranges from about 1.3 to 1.8 solar masses, depending on various factors. At this point, electrons are captured into the iron-peak nucleus, and the core becomes unstable. A violent supernova explosion occurs, or the star may collapse directly into a black hole.

In any of these ways, the star's life comes to an end. That's for the life of a solar-mass star, or the short and violent existence of a massive star. The least massive stars, the red dwarfs, will go on, as we've seen, practically for an eternity.

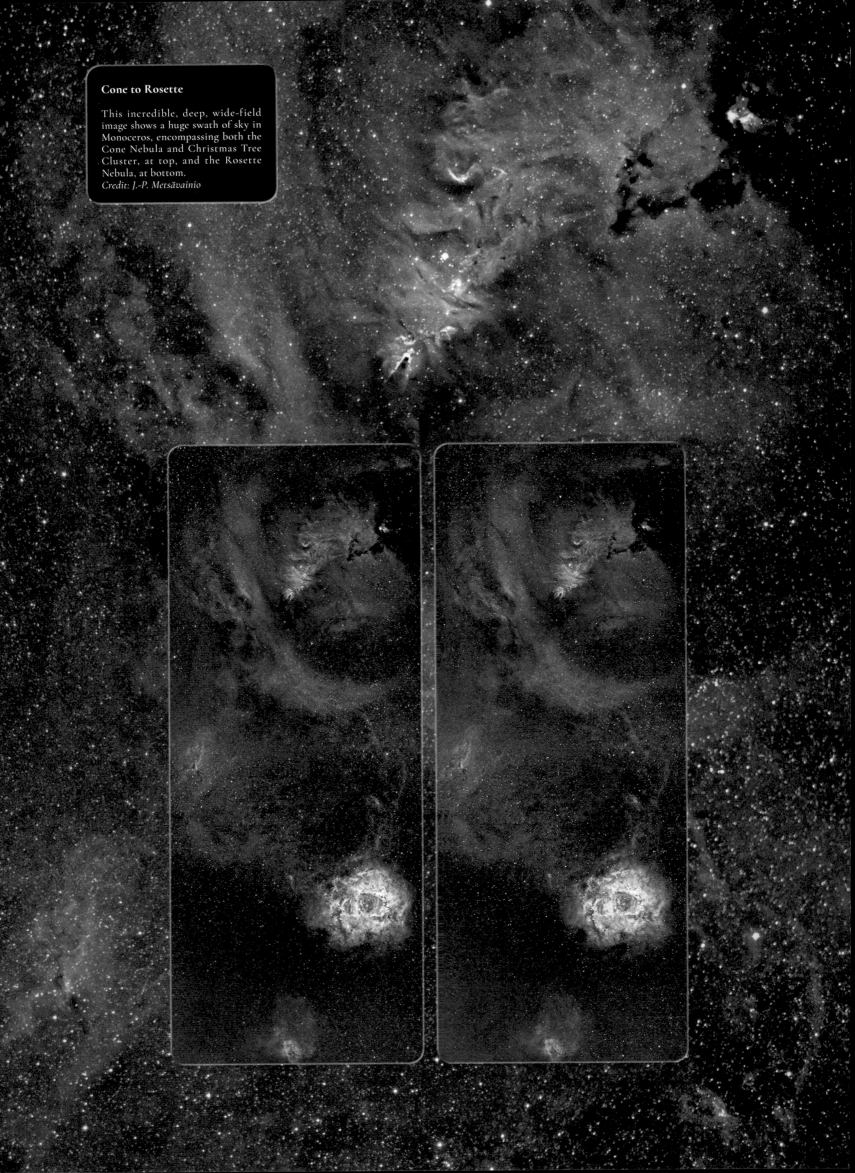

Cone to Rosette

This incredible, deep, wide-field image shows a huge swath of sky in Monoceros, encompassing both the Cone Nebula and Christmas Tree Cluster, at top, and the Rosette Nebula, at bottom.
Credit: J.-P. Metsävainio

INSIDE STELLAR BIRTHPLACES

5

As we've seen, the majority of emission nebulae are stellar nurseries, HII regions, that inhabit the thin disk of our galaxy (and other galaxies) and are busily collapsing by gravity in order to form protostars. They are often involved in the same localized area with reflection nebulae and with dark nebulae, depending on the content of the region and its geometry relative to our line-of-sight. Most often these nebulae produce open star clusters, the collections of newborn suns, that tidal forces disperse over long time periods as they orbit the galactic center.

We've briefly explored a few examples of the finest emission nebulae, including bright showpieces of the wintertime evening sky, like the Orion Nebula, the Rosette Nebula, and the Cone Nebula. We've looked at a couple of stellar examples lying in the deep southern sky, such as the Carina Nebula. We've even touched on some mammoth examples of HII regions that lie in other galaxies, such as the Tarantula Nebula (in the Large Magellanic Cloud) and NGC 604 (in the Triangulum Galaxy).

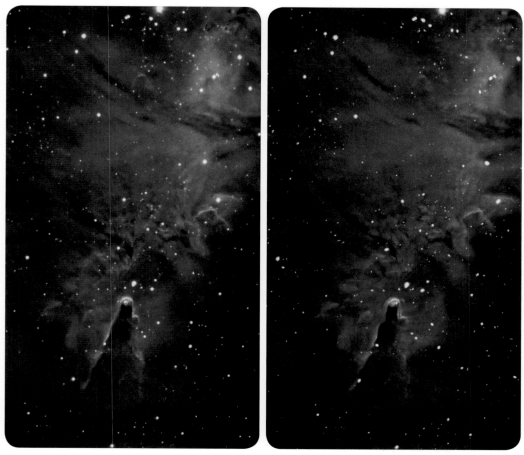

The Cone Nebula in 3-D

The Cone Nebula in Monoceros appears to float majestically, along with several streamers of dark nebulosity above it, in the foreground of this stereo image. The backdrop of bright nebulosity lies at a much greater distance.

How Does an HII Region Work?

Before we explore a range of other outstanding emission nebulae, let's refocus on what exactly is happening within these unique areas of the cosmos. Once they form inside giant molecular clouds, emission nebulae come to life when the most massive stars, usually O and B stars, are intensely hot enough to ionize them. Ionization happens when atoms or molecules gain negative or positive electrical charges as they gain or lose electrons. The resulting atom or molecule is called an ion. In the case of HII regions, it's the interaction with electromagnetic radiation that ionizes the atoms in the hydrogen gas cloud. When hot stars ionize the gas that surrounds them, they bombard the gas with ionizing photons, and the photons convert the gas into its HII form. So many HII regions can form within the spiral arms of galaxies that in some cases they can be used to trace the arms, which are peppered with regions of newborn stars.

The photons emitted from hot, massive stars sweep through the surrounding space and create an ionization front of glowing gas, which can move outward at supersonic speeds. The inverse square law applies, as it does everywhere in the universe, and so with increasing distance from the hot stars, the ionizing effect decreases in inverse proportion to (that is, divided by) the square of the distance. Newly ionized gas is energized, however, and so expands outward, causing its volume to expand, but eventually, far enough away from the stars, the ionization front slows to subsonic velocities. A mass collision is about to occur, however: when the expanding gas runs into the slowing ionized material, a shock front forms and shells, edges, bubbles, waves, and other formations light up as nebular material interacts.

Most HII regions last on the order of millions of years. The hot young stars being born within such clouds are ionizing the surrounding space, allowing the clouds to glow in the first place, but they are also bathing the surrounding region in radiation pressure, which eventually helps to drive away the lower-density gas. Normally, only a relatively small fraction of the gas within a cloud will transform into stars before the majority of the gas is driven away, finally to collapse into new stars, perhaps. The most massive stars within newborn clusters will also ultimately go supernova, after several million years, also helping to drive away some of the gas remaining within HII regions.

The stars that form in HII regions don't exactly do so out in the open. Believing that stars form in the denser, collapsed part of the interstellar medium, astronomers Bart Bok and his research partner E. F. Reilly spent part of the 1940s searching for very small, condensed dark clouds in and around bright nebulae. As we've described, they came to be known as Bok globules, and it was Bok's suggestion at a 1946 symposium that this was the location of stellar birth. By about 1990, just as astronomers were finally confirming the unambiguous existence of black holes, they also confirmed that Bok globules are the sites of star formation.

After the stars begin nuclear fusion and "turn on" as glowing balls of gas, they help to dissipate the remaining gas and dust in the globules. So, the sequence of star formation is that gas falls in by gravity to form clouds – HII regions, emission nebulae – which begin to glow as they are lit or energized by hot young stars forming within them. As they consume the gas, turning it into stars, or drive it away by radiation pressure or supernova explosions, the stars eventually obliterate most or all of the gas from which they were born. We are left with a newborn group of stars inside a fading and dissipating nebula, that breaks up and fades away as it circles the center of the galaxy. And then, when gravity gets its grip on gas somewhere else, the whole process begins again. And so goes stellar recycling.

Some of the smallest HII regions, so-called ultra-compact HII regions, are only about a light-year across, just about the size of our solar system. The largest HII regions, as we've seen – the Tarantula Nebula and NGC 604 are contenders – are more than 1,000 light-years across. In the 1930s the Danish astronomer Bengt Strömgren investigated the sizes of HII regions, and calculated the predicted sizes they could attain. Their bubbles of ionization, therefore, are called Strömgren spheres.

The density of the cloud and the energy coming from ionizing stars are the critical factors in determining how large the HII region can be. The densities can range from a million atoms per cubic centimeter in the ultra-compact HII regions to just a few atoms per cubic centimeter in the largest clouds. So, the range of masses of HII regions might span 100 to 100,000 solar masses of material. Their internal temperatures typically might be in the range of 10,000 kelvins, although they are complicated, multidimensional creatures, because HII regions can contain numerous ionizing stars, which complicates discussing their characteristics as a whole.

The composition of HII regions is approximately 90 percent hydrogen, with smaller quantities of helium and traces of heavier elements. The characteristic reddish color of the nebulae comes from the hydrogen-alpha emission line in their spectra, at 656.3 nanometers (a nanometer is one billionth of a meter). Astronomers call the heavier elements metals, and they are found in richer quantities closer to the galactic center, and dwindling amounts at greater distances from the core. Stars create heavier elements through fusion, as we've seen, and longer histories of star formation have taken place as one gets closer to the galaxy's core.

HOW WE SEE COLOR

As the reddish color of emission nebulae is so apparent in nearly all the photographs of them, you might wonder why a telescopic view of a nebula is so different. We see nebulae as fuzzy, gray-greenish areas of light in backyard telescopes because our eyes are not sensitive enough to see the characteristic hydrogen-alpha line very well, and because we don't see color unless it is bright. Our eyes have two types of detectors, rods and cones. The cones in the centers of our eyes are the color detectors, and they require pretty bright light. The rods, located more around the perimeter, are sensitive to faint light, but don't reveal colors very well. That's why, when peeking into a telescope, the technique of "averted vision" is so effective. Glancing slightly to the side of the field of view will make a faint nebula in the center of the eyepiece appear to brighten slightly, although its color is unaffected.

Some HII regions also are known to contain high-temperature plasmas (ionized gas that is highly electrically conductive), and they are so energized that they can emit X-rays. Some measure of X-ray emission has been recorded emanating from the Carina Nebula, the Orion Nebula, and the Omega Nebula. The cause may be ultra-hot O-type stars that are producing fierce stellar winds, with colliding winds creating shock fronts, or being accelerated by strong magnetic fields.

The Omega Nebula

One of the most spectacular nebulae in Sagittarius, lying in the direction of the galactic center, is the Omega Nebula (M17). The omega shape also appears like other figures, and so some have dubbed the object the Swan Nebula or Checkmark Nebula. This large star-forming region is about 5,000 light-years away and contains an embedded star cluster, NGC 6618. *Credit: Adam Block/Mount Lemmon SkyCenter/University of Arizona*

The Omega Nebula in X-rays

This image from the Chandra Observatory reveals hot gas, shown in red, flowing away from massive young stars in the center of M17, the Omega Nebula. The group of massive young stars firing this activity lie in the area shown in pink. *Credit: NASA/CXC/PSU/L. Townsley et al.*

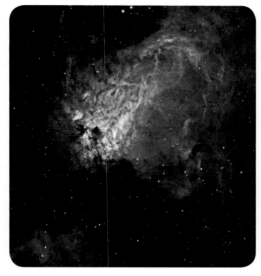
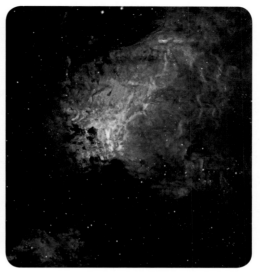

The Omega Nebula in 3-D

In three dimensions, the sharply defined Omega Nebula appears to unravel, its
brightest portions lying behind numerous layers of fainter nebulosity and dust.

Emission nebulae are not prevalent in all types of galaxies. They are abundant
in spiral, barred spiral, and irregular galaxies. They often result from galaxy colli-
sions, when interactions bring gas clouds into close proximity. But they are not
usually present in elliptical galaxies, which probably form through galaxy mergers,
and within which the gas clouds are perturbed, and don't often form new stars. In
colliding galaxies, starbursts can happen, however, in which rapid episodes of star
formation take place much faster than with the normal HII region method.

Such colliding or erupting galaxies that are undergoing accelerated episodes
of star formation are called "starburst galaxies". Notable examples include the
Cigar Galaxy (M82), the Antennae galaxies (NGC 4038–9), IC 10, Centaurus A,
and NGC 6946.

Additionally, a few rare observations have turned up HII regions existing outside
galaxies, but close to them. Such intergalactic HII regions may be the remnants of
galaxy interactions, the cosmic smash-ups that have left debris in their wake.

Despite the complexities involved with the physical structures of HII regions,
and the fact that they have limited lifetimes, the Milky Way Galaxy offers hundreds of good examples of these cosmic nurseries. As with all Milky Way objects,
they are best seen in the evening skies during the winter and the summer, when
times for skygazing are convenient and when the galaxy is well placed.

In the wintertime sky in the northern hemisphere, we've already seen the
incredible array of nebulae in Orion, mostly belonging to the Orion Molecular
Cloud, a relatively nearby complex of nebulosity and newborn stars. The adjacent constellation Monoceros also contains several large star-forming regions.
Also in the wintertime sky is the majestic constellation Auriga, and it contains a
variety of beautiful nebulae worth our attention.

Auriga is famous for three bright open star clusters belonging to the Messier
catalog of deep-sky objects, M36, M37, and M38. They are products of emission
nebulae now dispersed back into the interstellar medium. The most alluring nebulae in the constellation include the Flaming Star Nebula (IC 405), an emission
complex that also contains some reflection nebulosity. This object surrounds the
bluish variable star AE Aurigae, which normally shines at 6th magnitude, and lies
at a distance of some 1,500 light-years. The illuminating star, AE Aur, has a high

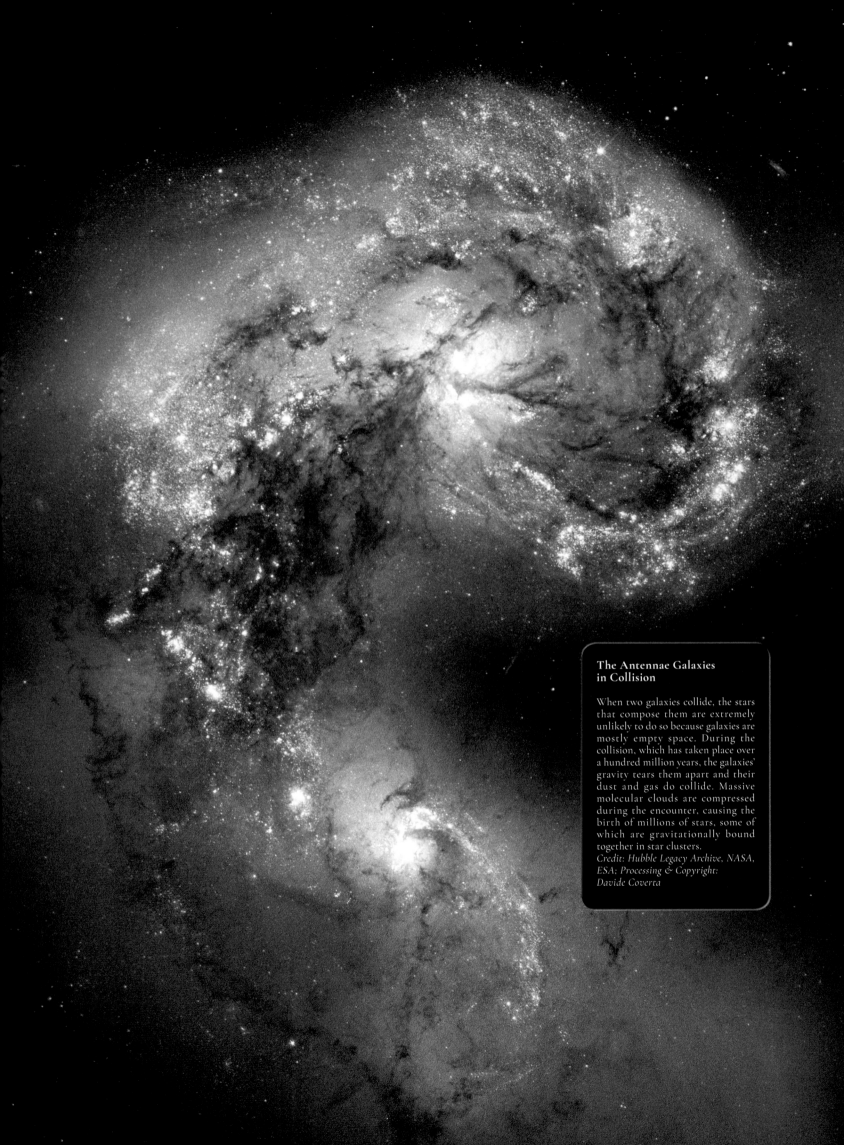

The Antennae Galaxies in Collision

When two galaxies collide, the stars that compose them are extremely unlikely to do so because galaxies are mostly empty space. During the collision, which has taken place over a hundred million years, the galaxies' gravity tears them apart and their dust and gas do collide. Massive molecular clouds are compressed during the encounter, causing the birth of millions of stars, some of which are gravitationally bound together in star clusters.
Credit: Hubble Legacy Archive, NASA, ESA; Processing & Copyright: Davide Coverta

Open Star Cluster M38

M38 is a rich open cluster of stars, each of which is about 200 million years old. Set in the disk
of our Milky Way Galaxy, M38 is young enough to contain many bright blue stars, although
its brightest star is a yellow giant shining 900 times brighter than our Sun. The cluster spans
roughly 25 light-years and lies about 4,000 light-years away. *Credit: NOAO, AURA, NSF*

proper motion across the sky, and may have originated from the Orion Molecular
Cloud, in the vicinity of Orion's Belt.

Lying nearby is another gorgeous emission nebula, IC 410, which is sometimes called the Tadpoles. This expansive star-forming region lies some 12,000
light-years away and contains the open cluster NGC 1893, which features stars
aged about four million years and younger. A curious feature within the nebula
includes two dense streamers of material trailing from the central, brightest part
of the nebula. These are sites of current star formation, and give the nebula its
curious name, as they are ten light-years long but look like tadpoles with bright
"heads" and wavy, fishlike "bodies."

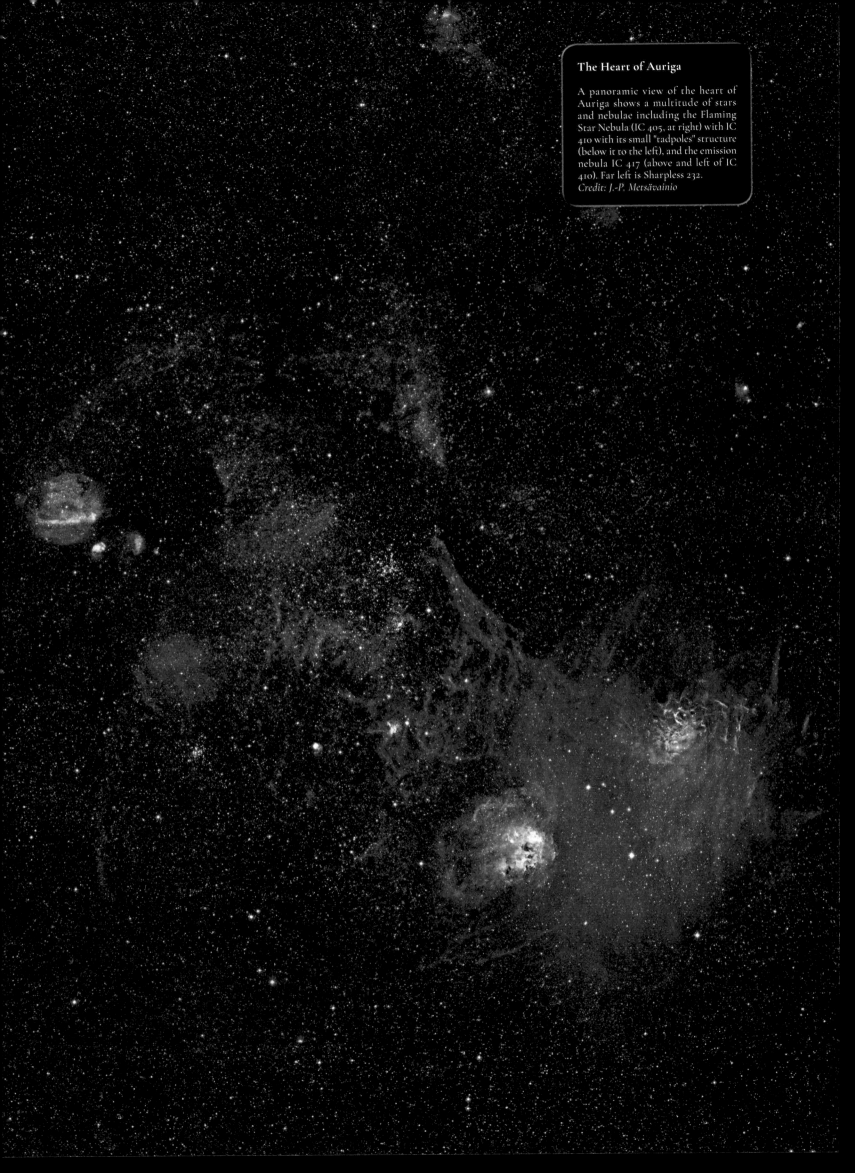

The Heart of Auriga

A panoramic view of the heart of Auriga shows a multitude of stars and nebulae including the Flaming Star Nebula (IC 405, at right) with IC 410 with its small "tadpoles" structure (below it to the left), and the emission nebula IC 417 (above and left of IC 410). Far left is Sharpless 232. *Credit: J.-P. Metsävainio*

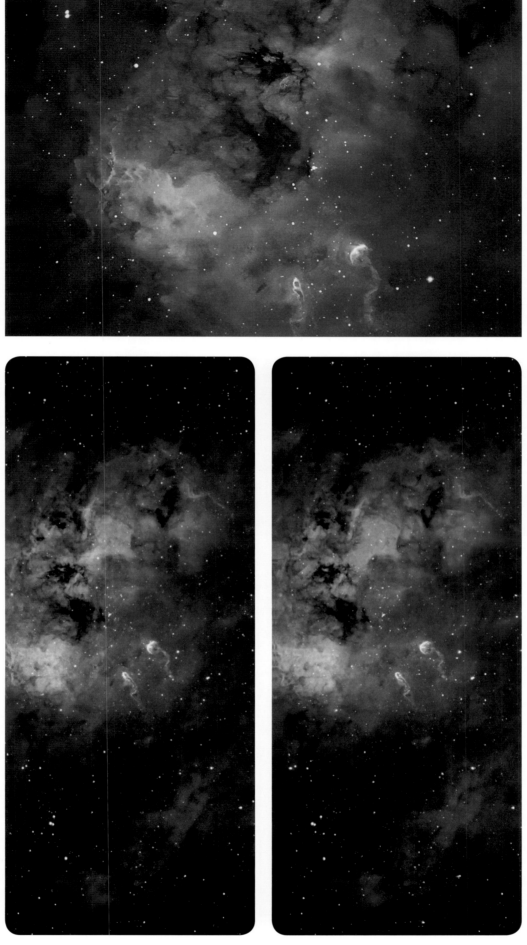

The Tadpoles Nebula in 3-D

The central area of IC 410 in Auriga, sometimes called the Tadpoles Nebula, shines in three dimensions. Dark nebulae float far in the foreground, and the tadpoles measure some ten light-years long and are regions of current star formation. The cluster embedded within IC 410 is NGC 1893, and the complex lies some 12,000 light-years away.

The California Nebula

The California Nebula (NGC 1499) is an emission nebula located in the constellation Perseus. It is so named because it appears to resemble the outline of the state of California. It lies at a distance of about 1,000 light years from Earth. Its fluorescence is due to excitation of hydrogen in the nebula by the nearby prodigiously energetic star, Xi Persei (also known as Menkib, seen above left of center). *Credit: J.-P. Metsävainio*

The winter sky in the northern hemisphere also holds one of the most remarkably shaped emission nebulae in the sky, the California Nebula (NGC 1499) in Perseus. Any image of this object reveals the obvious source of the nickname. Lying at a distance of some 1,000 light-years, the nebula is a shell that extends along our line of sight, but we see it as a fairly compact "ridge" of nebulosity, comprising the shape of the Golden State. The California Nebula is not only faint but is very large, extending across 2.5 degrees, equal to the width of five Full Moons placed side-by-side. So, in order to detect it in a telescope, one needs an exceptionally dark, moonless night, and a very low-power, wide-field scope.

Another remarkable emission nebula in the winter Milky Way is Thor's Helmet (NGC 2359), a complex of emission and reflection nebulosity featuring a central, smooth bubble of gas. Lying in Canis Major at a distance of about 12,000 light-years, Thor's Helmet received its nickname from a supposed resemblance to the Germanic, hammer-wielding god's headpiece. The bubble within this object is being "blown out" by an intensely hot Wolf-Rayet star, WR 7, which has 16 times the mass of the Sun and a surface temperature of 112,000 kelvins. This ionizes the surrounding gas and radiation pressure from the star carves the surrounding interstellar medium.

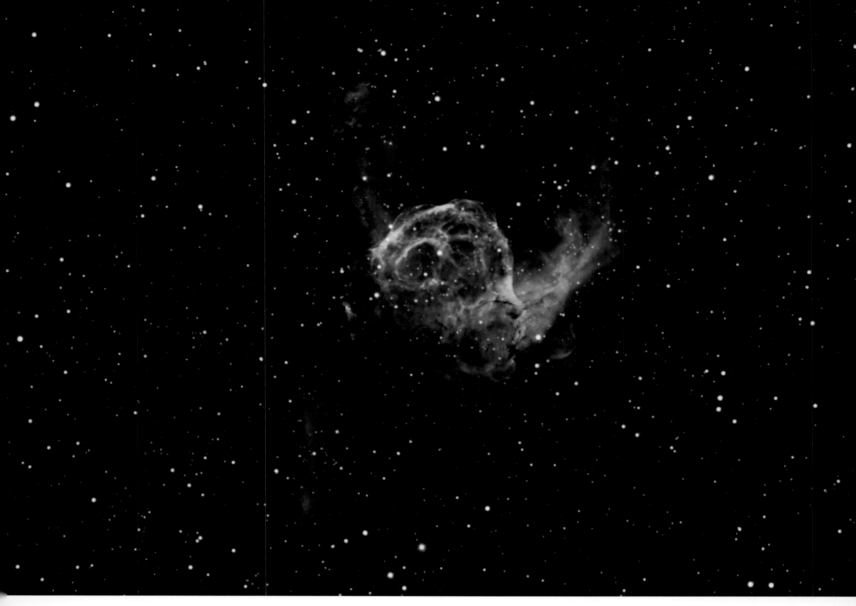

Thor's Helmet

The spectacular emission nebula Thor's Helmet (NGC 2359) appears in all its glory in this deep exposure. Lying in Canis Major at a distance of some 12,000 light-years, the nebula features a central bubble that is blown out by an intensely hot Wolf-Rayet star, ionizing the surrounding gas. *Credit: Bill Williams and Tony Hallas*

Also lurking in the wintertime Milky Way, faint to visual observers but a darling for cameras, is the Seagull Nebula (IC 2177). Lying at a distance of 3,700 light-years along the border of the constellations Canis Major and Monoceros, this sprawling region of emission and reflection nebulosity is churning out a young cluster of stars. A bright, detached portion of nebulosity, NGC 2327, is a round feature that is sometimes called the Seagull's Head. Two bright open clusters lie within; they are NGC 2335 and NGC 2343.

In the far northern sky lies the brilliant, "W" or "M" shaped group of stars known as Cassiopeia. High up within Cassiopeia lies the Heart Nebula (IC 1805), a faint target for visual observers but an astroimager's delight. The huge, heart-shaped cloud of emission nebulosity spans a large area and lies some 7,500 light-years away, lying in a rich Milky Way star field. Within this cloud lies a bright star cluster designated Melotte 15, and its stars illuminate the surrounding gas. Some of the stars in this group are 50 times the mass of the Sun, bathing the region in intense radiation.

The Seagull Nebula

A wide-field panoramic shot of the border between Canis Major and Monoceros shows the huge Seagull Nebula (IC 2177) throughout the right-hand side of the image. The bright, circular nebula near its upper right edge is van den Bergh 93. The small, bright nebula at the bottom of the Seagull is Cederblad 90. The cluster and wispy nebula at upper left are NGC 2353 and LBN 1036.
Credit: Joel Short

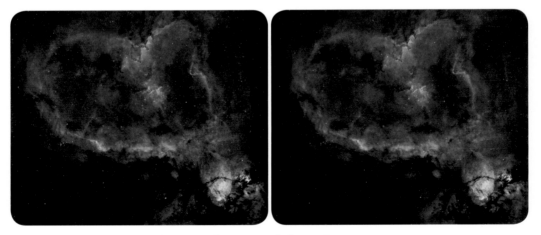

The Heart Nebula in 3-D

The Heart Nebula (IC 1805) in Cassiopeia explodes as a heart-shaped oval, with a spherical hot spot of star formation in its center, viewed in three dimensions.

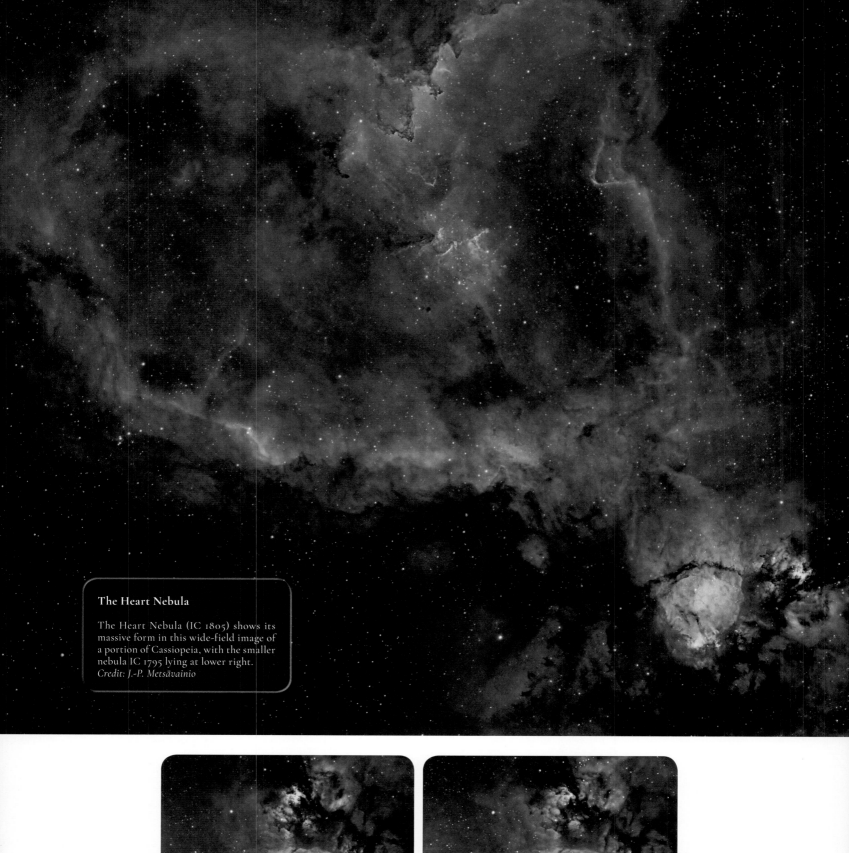

The Heart Nebula

The Heart Nebula (IC 1805) shows its massive form in this wide-field image of a portion of Cassiopeia, with the smaller nebula IC 1795 lying at lower right.
Credit: J.-P. Metsävainio

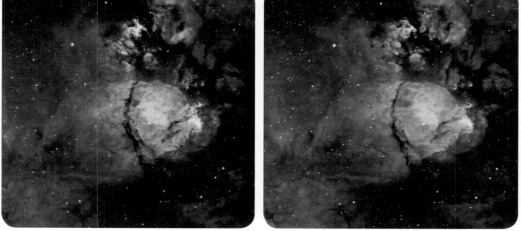

The Heart Nebula in 3-D (detail)

This detail blows up the far right portion of the Heart Nebula.

Melotte 15 in 3-D

A complex 3-D view of a small portion of the Heart Nebula (IC 1805) shows the bright star cluster Melotte 15 entangled in a rich and varied region of bright nebulosity. Dark towers appear to dominate this bewitching scene.

IC 1795

Lying on the perimeter of the expansive Heart Nebula (IC 1805) in Cassiopeia, oval-shaped IC 1795 features intricate zones of ionized gas and a prominent dark nebula crossing its face. *Credit: J.-P. Metsävainio*

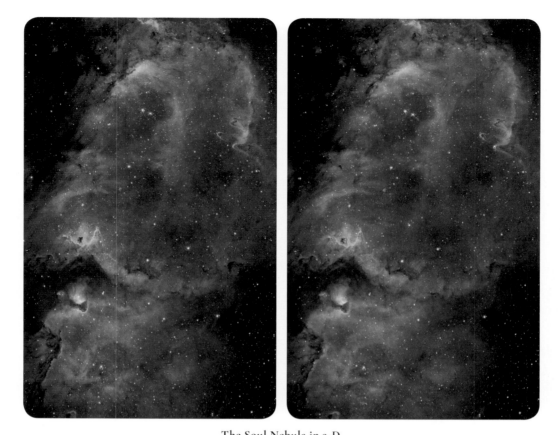

The Soul Nebula in 3-D

The Soul Nebula (IC 1848) transforms in three dimensions, showing a glowing cocoon with a few foreground stars floating far out in front.

Only a few degrees away in the sky is another large emission region, the Soul Nebula (IC 1848, sometimes called Westerhout 5). This big, double-lobed nebula is also faint for telescopic observers but a favorite astroimaging target, and lies at about the same distance as the Heart Nebula. The star cluster within, IC 1848, bathes the region in intense radiation, illuminating the gas. Strangely, just a short distance away from a line drawn between the two nebulae lie two very faint but famous galaxies, Maffei 1 and Maffei 2. These objects are relatively nearby galaxies but are dimmed so much by the dust in our galaxy that they went undiscovered until the 1960s.

The Soul Nebula

The Soul Nebula (IC 1848) lies near the Heart Nebula, also in a rich area of the Cassiopeia Milky Way, and contains two embedded star clusters, IC 1848 on the left side and Collinder 34 on the right.
Credit: J.-P. Metsävainio

Also lodged in the northern reaches of the sky, in Cassiopeia, is another beauty for the astroimager's camera, the Bubble Nebula (NGC 7635). Faint in a telescopic eyepiece, this beautiful object appears like a rectangular wedge of glowing light with a distinct, but faint, bubble of gas near its center. The ionizing radiation that is lighting up this region of star birth is SAO 20575, an intensely hot, 9th-magnitude star with a ferocious wind. This so-called Wolf-Rayet star is blowing a shell of material into the surrounding interstellar medium, creating the bubble shape, and the whole affair lies some 8,000 light-years away.

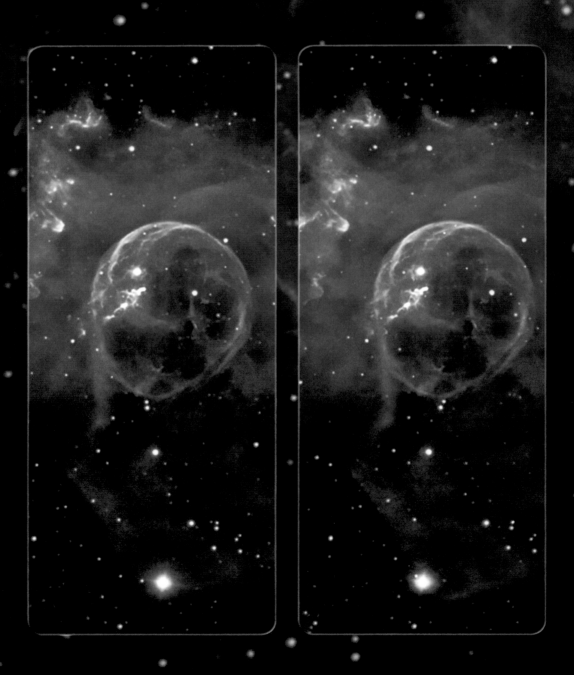

The Bubble Nebula

One of the sky's most breathtaking nebulae is certainly the Bubble Nebula (NGC 7635) in Cassiopeia. A so-called molecular cloud of reddish gas is visible throughout this image, along with many young stars. The Bubble Nebula itself appears like a perfect bubble, and excited into glowing, and expanding, by the hot central star, a blue-white star that is the bright star visible within the bubble.
Credit: J.-P. Metsävainio

NGC 7538

Lying near the more famous Bubble Nebula, emission nebula NGC 7538 in Cepheus is a roundish nebula with bright stars embedded that lies some 9,000 light-years away. *Credit: J.-P. Metsävainio*

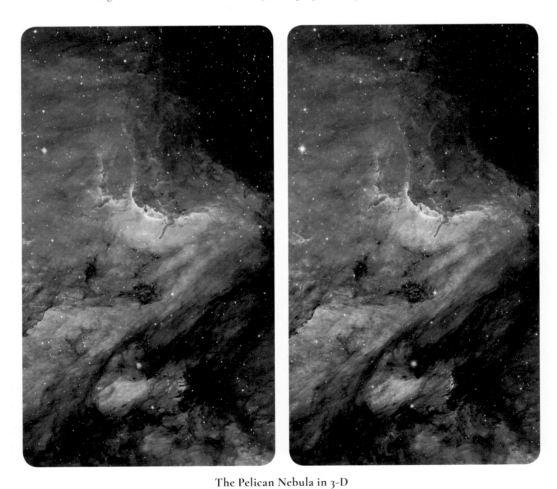

The Pelican Nebula in 3-D

In this brilliant portion of the Pelican Nebula in Cygnus, the pelican's beak and head appear to leap out against the dark nebulosity in the foreground.

North America and Pelican Nebulae

Two of the most distinctively shaped nebulae in the sky lie very close to the brightest star in Cygnus, Deneb. The North America Nebula (NGC 7000), left, is a huge emission nebula that is forming baby stars, some of which are speckled across its continental face. To its right is the Pelican Nebula (IC 5067/70), also named for its very distinctive shape. The North America Nebula is enormous – occupying a larger area of the night sky than a Full Moon – and can be seen with the naked eye from a dark sky site. These nebulae lie about 1,600 light-years away. *Credit: Adam Block /NOAO/AURA/NSF*

The northern sky also holds the majestic constellation Cygnus, smack dab in the plane of the Milky Way, which is studded with clusters and nebulae. The largest and brightest of the Cygnus gas clouds is the remarkable North America Nebula (NGC 7000), near the constellation's brightest star, Deneb. Also named for obvious reasons, the nebula traces out a pretty fair representation of not merely a state, but an entire continent.

The North America Nebula is huge, and may be faintly visible from a dark sky site with the naked eye alone. It makes a spectacular sight in binoculars or a wide-field telescope. The adjacent Pelican Nebula (IC 5067/70), also named for its obvious appearance, is part of the same large region of glowing nebulosity and is separated only by an intervening vein of dark nebulosity. These objects lie some 1,600 light-years away and constitute a very active star-forming region, containing open clusters NGC 6996, NGC 6997, and Collinder 428, and many dark nebulae, including Barnard 253 and Barnard 353.

So many amazing HII regions lie around us in a variety of directions relative to our humble galactic home. Yet it is no surprise that some of the very best and brightest are found in the constellation Sagittarius, which marks the direction toward the center of the galaxy.

One of the greatest in our sky is the Lagoon Nebula (M8), a huge star factory that graces a spot just north of the direction of the galactic center. This cloud of reddish gas spans some three times the width of the Full Moon and is visible as a misty spot of light with the naked eye. Consisting of two quite bright portions and a huge, wreath-shaped halo of gas, the object gets its name from a prominent dark nebula

The Lagoon Nebula

The Lagoon Nebula (M8) in Sagittarius is one of the most spectacular nebulae in the sky. This large star-forming region gets its name from the broad dust band that bisects it, and it lies some 4,000 light-years away. The rich, young star cluster within is NGC 6530, and the whole structure spans some 100 light-years. Several Bok globules, small dark nebulae collapsing to form protostars, lie within the nebula, and include Barnard 88 (top of the nebula), Barnard 296 (bottom of the nebula), and Barnard 89 (left side of the nebula).
Credit: R. Jay Gabany

that separates the bright portions. Lying some 4,000 light-years away, the nebula contains a bright open cluster, NGC 6530, full of ionizing stars. A series of dark nebulae help to define the cloud, including Barnard 88, Barnard 89, and Barnard 296.

Just north of the Lagoon Nebula, indeed within the same field of view of binoculars, is a complex object called the Trifid Nebula (M20). This complex of emission and reflection nebulosity contains a reddish, southern, circular half of emission nebulosity and a bluish, northern, circular half of reflection nebulosity. The ruddy emission nebula hosts a prominent, three-fold axis of dark nebulosity that separates the glow into three parts, giving rise to the name. The Trifid also lies about 4,000 light-years away.

This quick tour provides a look at just a few of the stellar examples of HII regions, the glowing emission nebulae scattered throughout the sky, giving us a look at the process of new stars entering the galaxy, slowly, and one by one.

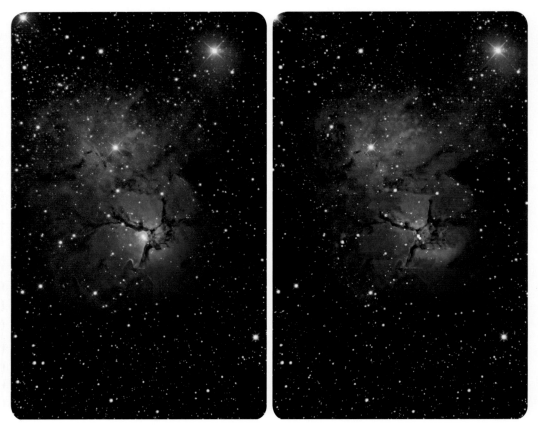

The Trifid Nebula in 3-D

In 3-D, the Trifid Nebula reveals its different components: a bright
central core is defined by a floating matrix of dark veins, and the bluish
reflection nebulosity receding into the background.

The Trifid Nebula

The Trifid Nebula (M20) in Sagittarius lies just north of the Lagoon Nebula, so close they can be seen in the same low-power telescopic field of view. The Trifid gets its name from the three-veined dark nebula that floats in front of its face. It lies some 4,000 light-years distant.
Credit: R. Jay Gabany

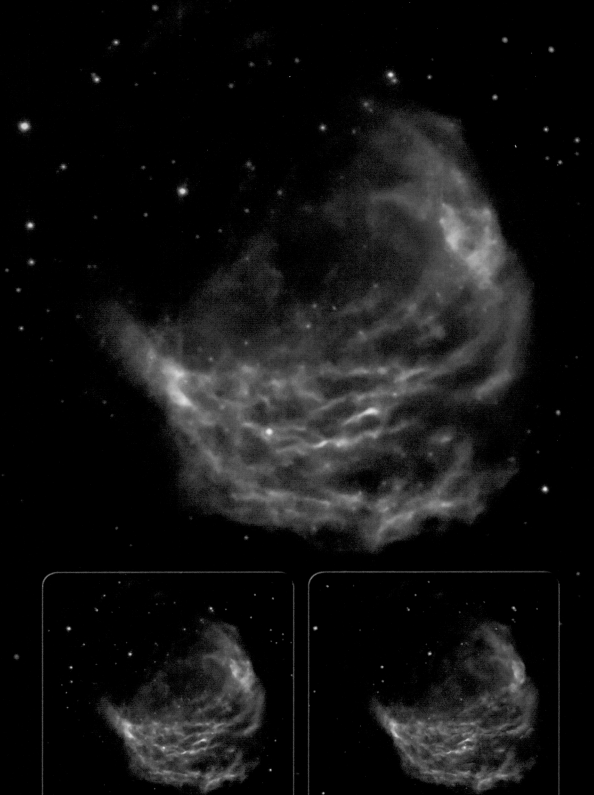

The Medusa Nebula

The Medusa Nebula is a large planetary nebula in the constellation Gemini on the border of Canis Minor. It is also known as Abell 21 and Sharpless 274 (Sh2-274). This nebula was thought to be a supernova remnant until early 1970; now it is known to be a planetary nebula at a distance of about 1,500 light-years. *Credit: J.-P. Metsävainio*

PLANETARY NEBULAE AND SUPERNOVA REMNANTS

Stars live and stars die, and they do so according to their masses. Low-mass stars can live so long that we don't know exactly how they will perish, as we've never seen one burn out. Dwarfs ought to quietly fuse their elements for a trillion years or more, and the universe is only 13.8 billion years old. For stars of intermediate mass, like the Sun, and for heavyweight stars, however, the story is pretty interesting and we know the details of how it happens.

Let's think about the Sun for a moment: Our own star, the source of all our energy and light, and the engine that makes our lives possible. We know it's about 4.6 billion years old, and that means it's about halfway through its life. Some five billion years from now, the Sun will exhaust itself of normal nuclear fuel and undergo a dramatic transformation, something akin to an ordinary caterpillar changing into a magnificent butterfly. Observers on planets in our part of the galaxy will not see a star in our spot anymore, but an expanding cloud of gas, a planetary nebula.

Astrophysicists have known that planetary nebulae are the remnants of dying, Sun-like stars for many decades, but the understanding of these transient objects underwent something of a revolution in the 1990s. At that time observations with the Hubble Space Telescope began to flow in, and the spectacular images returned by the orbiting observatory allowed astronomers to analyze the processes that unfold in making such a nebula.

When a Sun-like star nears the end of its life, exhausting the hydrogen needed to fuse, it becomes a red giant star. The star's core goes kaput and becomes an Earth-size ball of carbon with about half the Sun's original mass. The core no longer replenishes its heat, and it cools internally like an ember in a fireplace.

In the star's outer shells, gravity compresses hydrogen and helium and they burn ferociously, if briefly. For a time, the red giant star can be 150 times larger than its precursor, and 2,100 times more luminous. Carbon settles into the core, and not enough mass remains in the star to compress it to reignite fusion. So, the last chaotic bursts of helium burning fling layers of gas into space, resulting in the formation of a planetary nebula over the span of about a thousand years. These "belches" of gas produce a luminous bubble at typical outward speeds of 58,000 kilometers (36,000 miles) per hour, and we see these rings of gas, outer shells, in many planetary nebulae.

But the star's final helium flash is the defining moment for a planetary nebula. It goes off with great energy and defines the shape of the nebula to come. Inner structures form from this high-energy, higher-velocity spray of gas and dust. This flash often creates dark dust lanes, bright lobes, and other asymmetrical features, and the higher-velocity, second generation gas runs into the older gas and dust that are expanding more slowly, sculpting a variety of shapes. It is this interplay that makes each planetary nebula appear unique.

A planetary nebula lasts on average something like 50,000 years, so on cosmic timescales it is a relatively short-lived phenomenon. And enough Sun-like stars expire in our galaxy that we can see several hundred of these magnificent objects at any given time. It is the action between the layers of glowing gas, and the intensely hot, stripped down stellar remnant inside, the central star, that makes these objects so fascinating. These nebulae expand slowly and return their materials into the sea of the interstellar medium. And that leaves behind hot, bare remnants, white dwarf stars, that litter the galaxy and, after a billion years, fade into virtual nonexistence, too faint to see from appreciable distances.

After the dying star initiates its planetary nebula phase, it takes perhaps a thousand years for the nebula to become visible. The gas needs to expand into a large enough space to show up, and the star needs to slough off its outer layers, exposing a brutally hot layer that can help to ionize the expanding gas. This intensely radiative remnant reaches temperatures of about 55,000 degrees. Before the stage of intense ionization from the star, the nebula glows dimly from dust particles that reflect light toward us, like a reflection nebula.

The process of ionization is what really brings on the show. With the star glowing at such extreme temperatures, its ultraviolet photons strip electrons from neutral atoms. This is the ionization process, and the resulting glow makes the nebula visible. The planetary nebula shines in spectral lines of hydrogen, helium, and other elements, and it can be observed in telescopes, and captured by backyard astronomers armed with CCD cameras (see Glossary) on their telescopes.

The amazing variety of shapes and sizes and brightnesses captured by the Hubble Space Telescope has given astronomers plenty to work with. We certainly understand the mechanics of planetary nebulae far better than we did a generation ago, but there are still major unanswered questions about how they form and what details in their structures suggest.

Some of the questions start even in the protoplanetary stage, when a nebula is beginning to form. Astronomers believe that dust should be flowing outward from the central star, like the tail of a comet pushed in a particular direction by the solar wind. With a spherical star, the dust should be radiating out uniformly. But that's not what they see in protoplanetary nebulae. Astronomers have also found that the outward gas flow is much stronger than could be explained by outward pressure from the star.

Abell 61

The planetary nebula Abell 61 in Cygnus may appear to be a spacecraft prop from the original *Star Trek* series, but it is an old, highly evolved object. Lying about 4,500 light-years away, it is some 22,000 years old and contains a binary star in its center.
Credit: Don Goldman

Dengel-Hartl 5

Lying in the constellation Cepheus, near the long reflection nebula van den Bergh 152, is the faint, strangely-shaped planetary nebula Dengl-Hartl 5. This object is so faint that it was discovered only in 1979, and it has expanded in an irregular way into the surrounding space.
Credit: Don Goldman

Ellis-Grayson-Bond 6

The peculiar and old planetary nebula Ellis-Grayson-Bond 6 lies in Leo and appears to be a weathered, weary shell of gas. It is some 1,500 light-years distant and is a faint object for observers and astroimagers.

Credit: Don Goldman

Heckathorn-Fesen-Gull 1

"Old soldiers never die, they simply fade away," said Douglas MacArthur. Well, old planetary nebulae fade away too, and that's exactly what Heckathorn-Fesen-Gull 1 is doing. Located in Cassiopeia and some 1,500 light-years away, this nebula is more than 10,000 years old and is fading into the interstellar medium.

Credit: Don Goldman

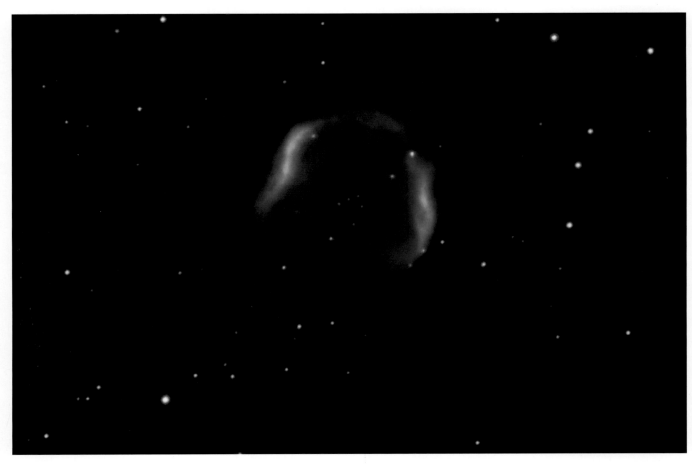

Jones 1

One of the faintest and strangest planetary nebulae known is Jones 1, a bluish shell of
gas in Pegasus. Lying some 2,300 light-years distant, the nebula is too faint for most
visual observers, and even a difficult target for astroimagers to record. *J.-P. Metsävainio*

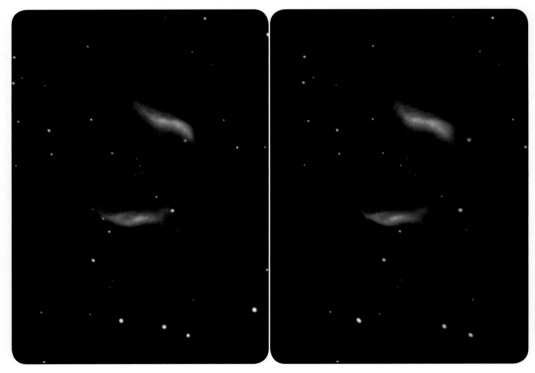

Jones 1 in 3-D

The tortured shell of planetary nebula Jones 1 appears multidimensional in this view, with its tiny central star
floating within the sphere, and the other faint stars near it receding into the background.

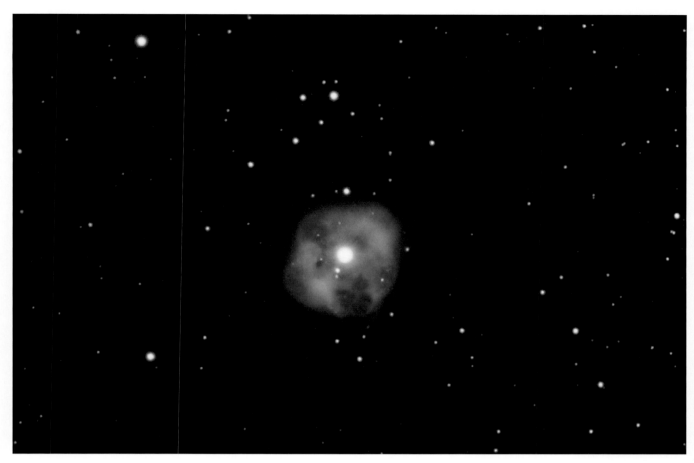

NGC 1514

Planetary nebula NGC 1514 in Taurus was discovered in 1790 by William Herschel. It seems to show a twisted, tortured appearance within, surrounding a bright central star. The nebula lies some 2,200 light-years away. *Credit: Dietmar Hager*

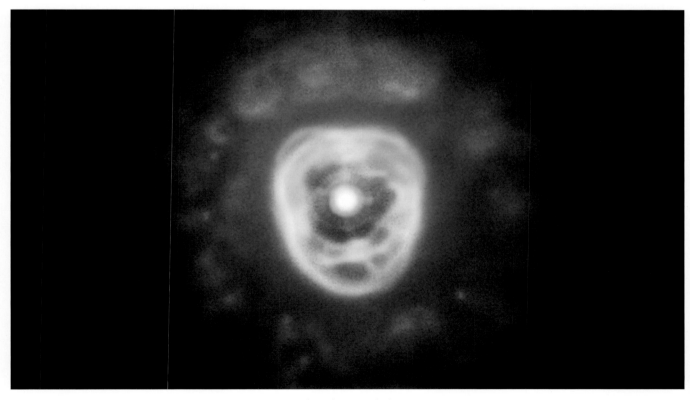

The Eskimo Nebula

The Eskimo Nebula (NGC 2392) in Gemini is one of the most universally recognized planetaries. The nebula's light seems to form a "parka" hood. It is a bipolar nebula that has produced two distinct shells, a brighter, inner shell, and the faint, outlying hood. The Eskimo lies some 6,500 light-years away and is bright enough to see in small telescopes. *Credit: Peter and Suzie Erickson/Adam Block/NOAO/AURA/NSF*

NGC 4361

Located in the southern constellation Corvus, the bright planetary nebula NGC 4361 presents a multi-shell structure that shows up well visually in large backyard telescopes. Lying some 3,400 light-years distant, the nebula features a bright central star.
Credit: Adam Block/Mount Lemmon SkyCenter/University of Arizona

The Cat's Eye Nebula

The Cat's Eye Nebula (NGC 6543) in Draco is a well-known object for both observers and casual science enthusiasts. Its multilayered rings of light are deep within the center of this image, which records huge, older, fainter shells of material around it. The nebula lies some 3,300 light-years away.
Credit: J.-P. Metsävainio

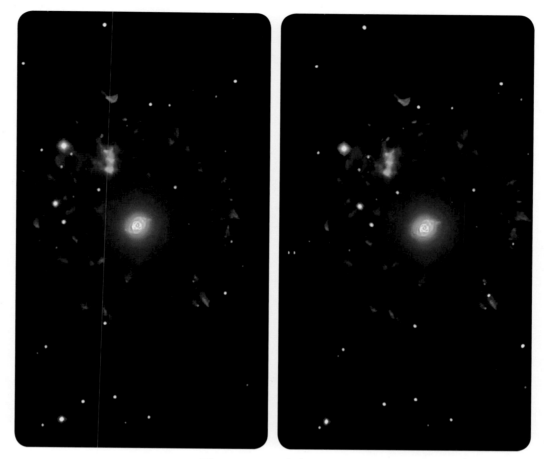

The Cat's Eye Nebula in 3-D

Bewitchingly beautiful and starkly isolated in space in 3-D, the Cat's Eye Nebula (NGC 6543) appears like a sculpture the size of an entire solar system. In its center lies the bright "cat's eye," while an enormous, faint outer shell punctuated with blobs of nebulosity surrounds it.

NGC 6894

NGC 6894 is a planetary nebula in Cygnus with an unusual shell shape and multiple bubbles that are highlighted by strands of nebulosity running throughout. The nebula lies some 5,400 light-years away and is bright enough to be observed with backyard telescopes. *Credit: Adam Block/NOAO/AURA/NSF*

NGC 7048

Another distinctive planetary nebula in Cygnus, NGC 7048, shows a ring peppered with faint stars in a very rich star field. Slightly elliptical in shape, it lies some 5,300 light-years away.
Credit: Richard Robinson/Beverly Erdman/Adam Block/NOAO/AURA/NSF

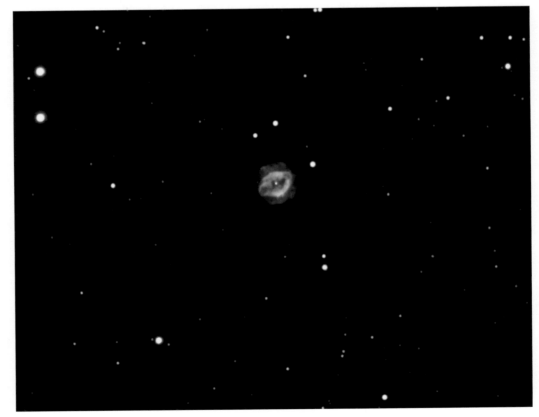

NGC 7354

The bright blue planetary nebula NGC 7354 in Cepheus presents an elliptical inner shell surrounded by a very faint outer halo of older gas. This peculiar nebula lies some 4,200 light-years away.
Credit: Adam Block/Mount Lemmon SkyCenter/University of Arizona

NGC 7662

The beautifully complex planetary nebula NGC 7662 in Andromeda, sometimes called the Blue Snowball Nebula, shows inner rings surrounded by an immense outer halo of bluish light. It lies some 5,600 light-years away and is easily visible in small telescopes. *Credit: Derek Santiago*

The Red Rectangle Nebula

A protoplanetary nebula, the so-called Red Rectangle (HD 44179) in Monoceros is in the earliest stages of a Sun-like star's death. This unique bipolar nebula shows X-shaped spikes of material flowing from the aging star, a binary system. The object, lying 2,300 light-years away, is laden with dust. *Credit: Adam Block/Mount Lemmon SkyCenter/University of Arizona*

Planetary Nebula in Lynx

PuWe1 is a very large and diffuse planetary nebula in the constellation Lynx. Total exposure time to capture this picture was around 23 hours. There is a hint of outer halo visible in this image.
Credit: J.-P. Metsävainio

BLOWING IN THE STELLAR WIND

After the nebula forms, we know that a stellar wind sets up the initial formation of the planetary nebula, and then a fast wind – up to 560,000 kilometers (about 360,000 miles) per hour – runs into the slower wind, which had been ejected in an earlier phase. The fast wind cooks everything in its way, creating the appearance of a cavity between the central star and the glowing shell we see. But the cavity is not really empty – it just appears that way because it's less dense.

Outside the internal cavity, a thin rim defines the glowing nebula that we see from afar. Outside the rim, the material expanding slowly from the slower, earlier wind is not yet overrun by the fast wind. When this happens, it will pop the bubble and empty the contents into the interstellar medium. Lobes of gas, large bubbles, and other formations can occur during this transformation.

As the central star increases its temperature and begins to ionize the surrounding gas, the wind strips layers from the star, exposing increasingly hotter regions, and the star's surface becomes more bluish in color. The star's surface temperature can eventually reach about 180,000 degrees. The ionization that results causes the expanding gas to fluoresce, and we can see it from afar.

Buoyed by the incredible imaging from the Hubble Space Telescope, astronomers have made a good start in understanding the shapes of planetary nebulae, but lots of work is left to be done. The basic shapes are round, elliptical, and butterfly-shaped. For some reason, many protoplanetary nebulae are elliptically shaped, and have prominent dust lanes. But only ten percent of mature planetary nebulae are bipolar, with a strongly elliptical shape. Why protoplanetary nebulae evolve into mature nebulae with somewhat different shapes remains a mystery.

The shapes of planetaries are highly symmetrical. The cause of this in some cases may be from double star systems wherein a companion tugs on the outer layers of a red giant, or maybe a giant consumes its companion as it enlarges. The double star scenario could cause a spiral pattern in which material moves out in symmetrical form. In the case of the swallowing of a companion, perhaps the smaller star spirals in in a symmetrical way, forming a disk that shapes the nebula.

In the minds of most astronomers, mergers of binary stars are a popular way to account for many observed shapes of planetaries. But magnetic fields created by the convection happening in the host stars could also be important. During the helium flash of a red giant, the intense heat causes material in the star to rise quickly, and the star could then eject magnetized gas, shaping it as it's still attached to the star's surface. This could also explain some of the shapes astronomers observe.

Although a great deal of research still needs to be done to understand the details of how planetary nebulae form, and how they are shaped, we know the basic story. The beautiful, symmetrical shapes of these transient glows, marking the demise of Sun-like stars, pepper the galaxy. These objects are bright enough to see in our backyard telescopes. They offer great beauty, and a forward look at what will happen to our Sun, and our solar system, some five billion years from now.

THE DEMISE OF MASSIVE STARS

For massive stars, those of about eight solar masses and above, the process of ending their lives is far more dramatic. As they deplete their stockpiles of hydrogen and other elements needed for fusion, these stars also grow and cool at first. They do not brighten as ferociously as smaller stars during this phase, but they evolve toward becoming red supergiants. At some point their cores cannot sustain themselves by outward pressure and they move toward collapsing into neutron stars or black holes.

The smallest and densest suns, neutron stars typically measure a mere ten kilometers across and contain little more than one solar mass of material. Thus, their density is so great that a teaspoon of its material would weigh a billion tons. If you squeezed Earth into the density of a neutron star, it would span a mere 300 meters. These bizarre stellar remnants get their name from the fact that they contain almost exclusively neutrons, subatomic particles with no electrical charge.

Another end result of a massive star is a stellar black hole. These fantastic creatures are small regions of space wherein the gravitational attraction is so strong that nothing, not even light, can escape. First hypothesized by the English natural philosopher John Michell in 1783 as "dark stars," astronomers lacked concrete evidence for them until 1990, when it became clear that Cygnus X-1, a strong X-ray source, had to be a stellar black hole. In the meantime, Albert Einstein added vast theoretical support for the idea that these regions of spacetime should exist.

But to get to a neutron star or a black hole, first massive stars must experience one of the most violent events in the universe. As massive stars deplete their hydrogen, their cores grow hotter and denser, and they begin to fuse carbon and heavier elements. After they fuse the available carbon, their core masses are about 2½-times that of the Sun, and they undergo neon burning. For stars of about 8 to 12 times the mass of the Sun, this results in runaway fusion and the star is destined to explode as a supernova.

For more massive stars yet, the oxygen burning phase leads to the fusion of neon and silicon, and this leaves a core of mostly iron. This iron-rich core grows until it reaches a critical mass, between one and two solar masses, and then electrons are captured into the iron nuclei in the core and the star can no longer support itself. The star also explodes violently in a supernova blast.

Supernovae are pretty rare and extraordinarily dramatic events. For a brief time, they can outshine the entire rest of the galaxy in which they reside. They expel several solar masses of material at velocities close to the speed of light, and so they add material from the dying star back into the interstellar medium, producing what astronomers call a supernova remnant. We can see these supernova remnants for timescales of several tens of thousands of years, and sometimes longer. Interesting examples of them lie scattered across the sky.

In our Milky Way Galaxy, only three naked-eye supernovae have appeared in the last millennium: The Crab Supernova in 1054, Tycho's Supernova in 1572, and Kepler's Supernova in 1604. The most recent bright supernova to appear was Supernova 1987A, which was visible to the naked eye, but it appeared in the Large Magellanic Cloud, the Milky Way's largest satellite galaxy.

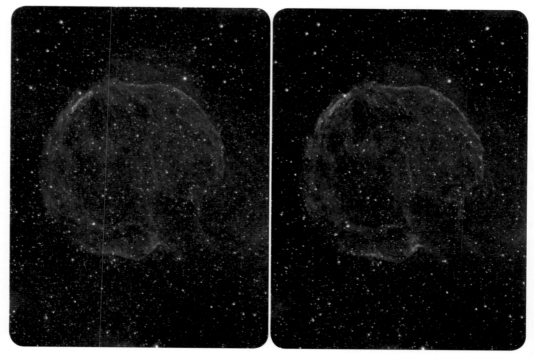

CTB1 in 3-D

In three dimensions, the ghostly supernova remnant CTB1 (Abell 85) appears as a wisp of a shell, floating and barely visible in the flutter of Milky Way stars and nebulae mostly behind it.

CTB1

Originally believed to be a planetary nebula, CTB1 (Abell 85) is now known to be a rather distant supernova remnant, the shell of expelled gas from a massive star's death. Lying some 10,000 light-years away in the constellation Cygnus, this glowing shell of gas will dissipate into the interstellar medium in the coming tens of thousands of years.
Credit: J. P. Metsävainio

The Jellyfish Nebula

Sometimes called the Jellyfish Nebula, IC 443 is a supernova remnant in Gemini, the remains of a star that exploded perhaps 10,000 years ago. Lying some 5,000 light-years away, the nebula lies close to the naked-eye star Eta Geminorum (at right) in our sky. *Credit: J.-P. Metsävainio*

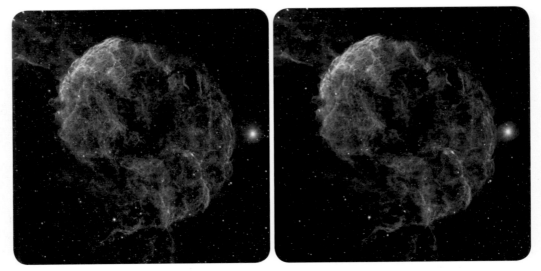

The Jellyfish Nebula in 3-D

The Jellyfish Nebula (IC 443) comes alive in stereo, showing a more understandable supernova remnant bubble, with a prominent cloud of dust floating way out in front, helping to shape it. The bright star Eta Geminorum is a dramatic foreground object.

TYPES OF SUPERNOVAE

Supernova remnants can result from a couple of different processes. A Type Ia supernova results from a pair of stars orbiting each other. The more massive star grows old and becomes a giant, delivering gas onto the smaller star, causing it to expand. The smaller star and the core of the heavier star spiral inward, producing a common envelope that is blown away by radiation pressure. The giant's core then collapses into a white dwarf star. The aging companion then flows material onto the white dwarf, which reaches a critical limit, and violently explodes, producing the supernova.

These Type Ia supernovae are very well understood and they have absolutely standard brightnesses, behaving in a systematic and predictable manner. This allows them to be one of the most valuable tools in an astrophysicist's box, as they can be used to calculate the distances to galaxies in which they appear. They are thus referred to as "standard candles."

So-called Type Ib and Ic supernovae result from stars that have lost most of their outer envelopes due to strong stellar winds or interaction with other stars. They are massive stars that simply undergo core collapse. These unusual stars are called Wolf-Rayet stars after the astronomers who studied them carefully, Charles Wolf and Georges Rayet at the Paris Observatory in the 1860s. Type Ib supernovae result from Wolf-Rayet stars that have helium in their atmospheres. Type Ic supernovae are devoid of helium in their atmospheres.

Aside from exotic stars with high winds or binary systems with interplay between their suns, most massive stars that go supernova simply explode like a bomb. They are termed Type II supernovae. These stars feature late-stage burning of elements in their cores and atmospheres like the layers of an onion, and when their cores collapse, they have a hydrogen envelope, and they explode.

Supernovae transform their galactic vicinity, and not in good ways. They blast the space surrounding them with high-energy, sterilizing radiation that could kill any life on nearby planets. So, we'll hope that no local stars go supernova soon. That is exceptionally unlikely, fortunately. The nearest candidates for supernovae include the star Spica in Virgo, 250 light-years away; Antares in Scorpius, at 554 light-years; Betelgeuse in Orion, at 720 light-years; Deneb in Cygnus at 1,410 light-years; Mu Cephei, at 5,900 light-years; and Eta Carinae, at 8,630 light-years.

Long-term dangers aside, supernovae have left some beautiful remnants for us to observe, as they scatter material back into the interstellar medium.

TYPES OF SUPERNOVAE

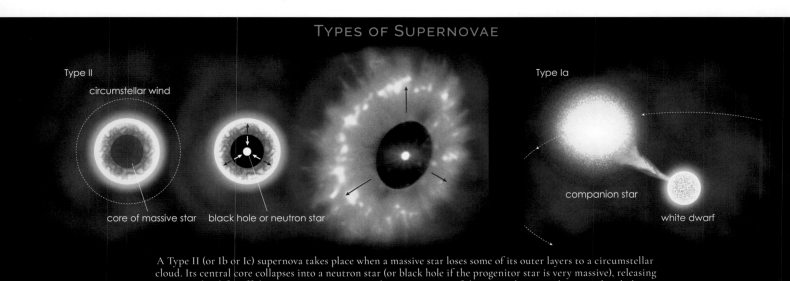

A Type II (or Ib or Ic) supernova takes place when a massive star loses some of its outer layers to a circumstellar cloud. Its central core collapses into a neutron star (or black hole if the progenitor star is very massive), releasing energy that lifts off the progenitor star's atmosphere. Fragments of the atmosphere streak outward and plow into the surrounding space, outlining a spherical cavity that becomes visible as a supernova remnant. A Type Ia supernova is the explosion of a white dwarf star that reignites nuclear fusion either by accretion of material from a nearby star or the merger of two stars. *Credit: James Symonds*

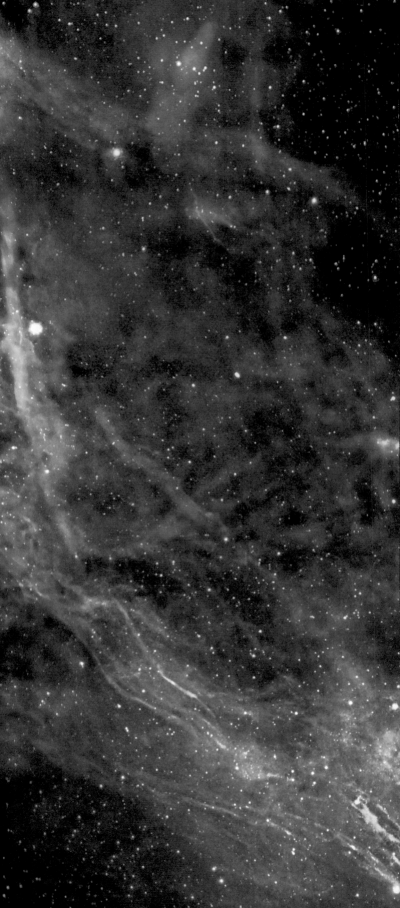

THE VEIL NEBULA

One of the most spectacular examples lies in the bright northern constellation Cygnus, and is known as the Veil Nebula (NGC 6960, NGC 6979, and NGC 6992–5). The Veil is a large curved shell of nebulosity with several of its areas strongly ionized, and they glow, marking the boundary between the expanding gas and the interstellar medium. This magnificent object is quite faint, but portions of it can be glimpsed in moderate-size telescopes and it can be imaged relatively easily. The nebula resulted from a 20-solar-mass star that exploded about 8,000 years ago.

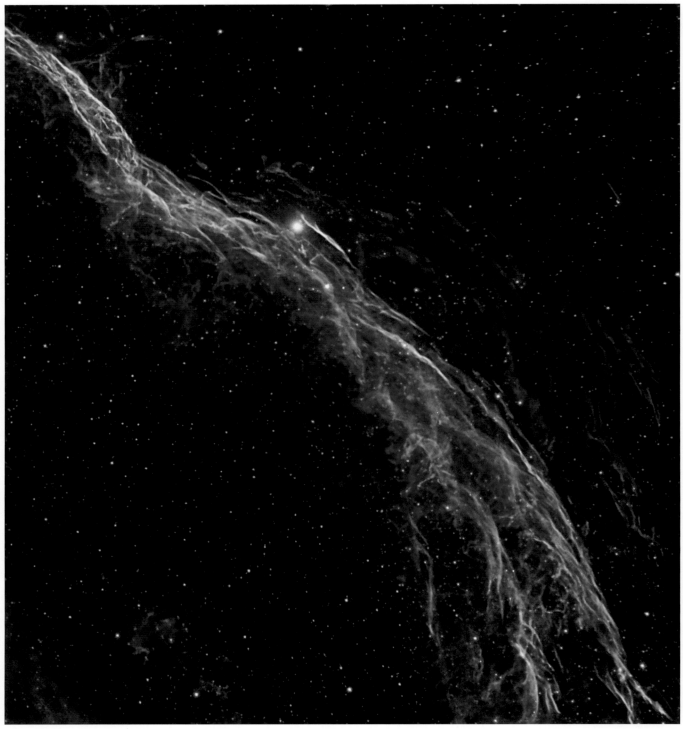

The Witch's Broom Nebula

NGC 6960, the westernmost portion of the Veil Nebula, which despite its 1,470 light-year distance covers over five times the diameter of a Full Moon. *J.-P. Metsävainio*

The Eastern Veil Nebula

The easternmost portion of the Veil Nebula, NGC 6992–5, contains a bright,
"twisted rope" structure that is visible in large backyard telescopes. It dramatically
stands out in a very rich Milky Way starfield. *Credit: J.-P. Metsävainio*

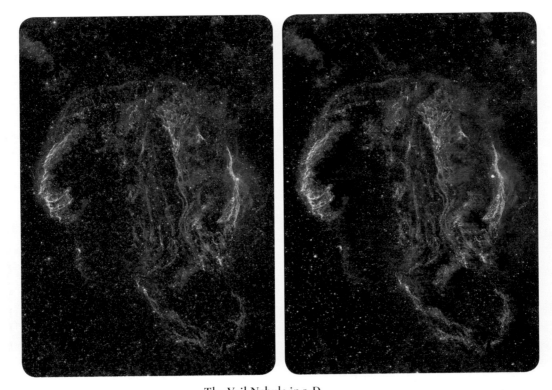

The Veil Nebula in 3-D

Viewed in stereo, the bulbous shape of the entire Veil Nebula comes alive, and we see the bright nebulae NGC 6960 and NGC 6992–5 as "hot spots".

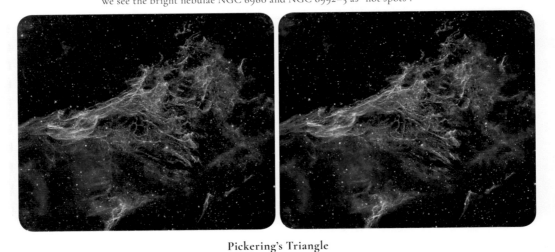

Pickering's Triangle

The portion of the Veil Nebula known as Pickering's Triangle appears in stereo like a flattened, hollow tube, with ropy structures wrapping around all of its sides.

Discovered by William Herschel in 1784, the Veil lies some 1,470 light-years away and is characterized in images by beautiful, rope-like strands and filaments of glowing gas. At the Veil's distance, its size on the sky translates to a physical diameter of about 80 light-years. The western side of the nebula, NGC 6960, appears to pass through the bright star 52 Cygni and is sometimes called the Witch's Broom. The eastern half is NGC 6992–5, and shows the most complex, "twisted," ropy structure. A relatively bright portion between the two sides, NGC 6979, is called Pickering's Triangle in honor of Harvard astronomer E. C. Pickering, although Pickering's assistant Williamina Fleming actually discovered the nebula.

Perhaps the easiest supernova remnant to observe with small telescopes is the Crab Nebula (M1) in Taurus (see chapter 3). It is bright enough to see in small telescopes and lies some 6,500 light-years away, its tendrils of gas (that give rise to the nickname) expanding within its 5.5-light-year diameter. Deep within the Crab lies a pulsar, a rapidly-rotating neutron star.

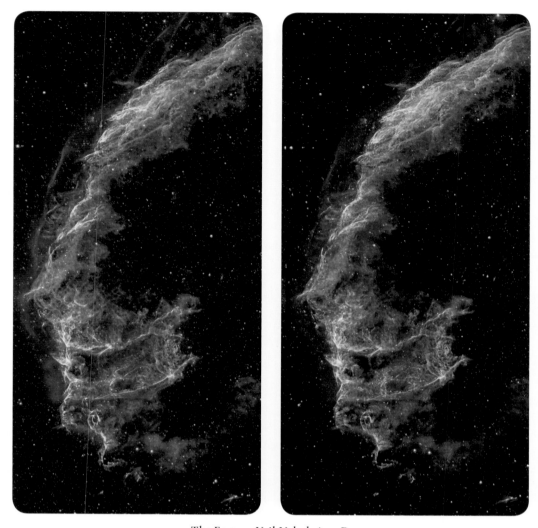

The Eastern Veil Nebula in 3-D

In stereo this view of the NGC 6995 component of the Veil Nebula comes alive, making it one of the most beautiful and complex objects in the sky.

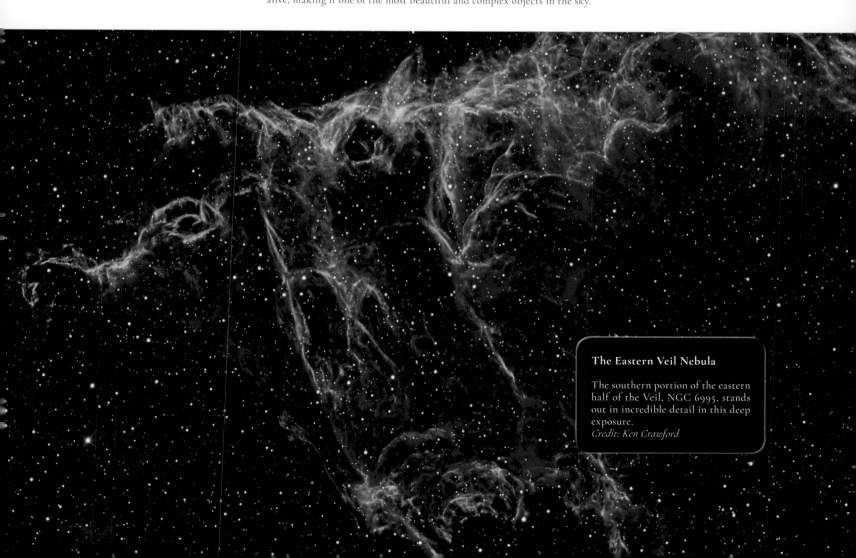

The Eastern Veil Nebula

The southern portion of the eastern half of the Veil, NGC 6995, stands out in incredible detail in this deep exposure.
Credit: Ken Crawford

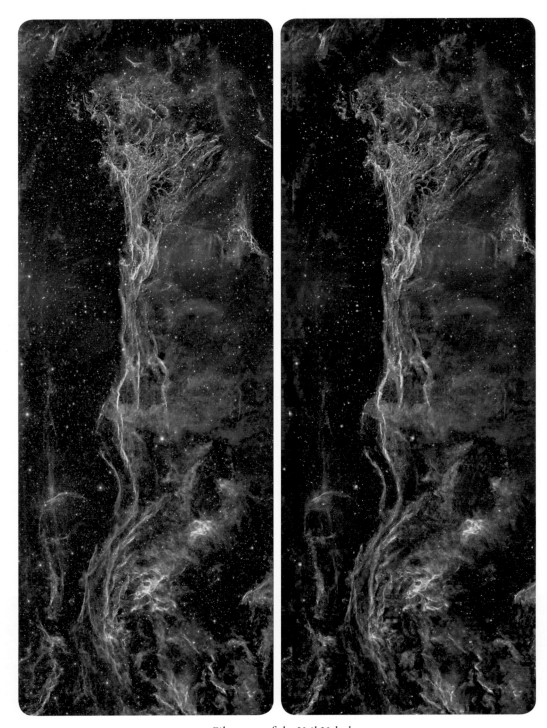

Filaments of the Veil Nebula

This new image is a four-panel mosaic showing the complex, filament-like structures in the Veil. Total exposure time is around 35 hours.

Many other supernova remnants lie scattered around the sky, although most are quite faint. In the southern hemisphere, the Vela Supernova Remnant is a large and notable example. Lying only 800 light-years away, this large cloud appears to be a lace-like matrix of glowing tendrils over a large area of sky.

Back in the north, the supernova remnant known as Simeis 147, sometimes called the Spaghetti Nebula, offers another intricate array of faint tendrils of interlaced gas. Lying along the border of the constellations Auriga and Taurus, this big object lies some 3,000 light-years away.

The Pencil Nebula

The curious object known as the Pencil Nebula (NGC 2736) is an oddity, a part of the much larger Vela Supernova Remnant. Lying in the deep southern sky, this object is a portion of the gas violently erupted when a massive star blew itself apart at the end of its normal life, around 8,000 years ago. Lying some 800 light-years away, the supernova remnant will gradually disperse into the interstellar medium, eventually finding itself recycled into new generations of stars.
Credit: Ken Crawford

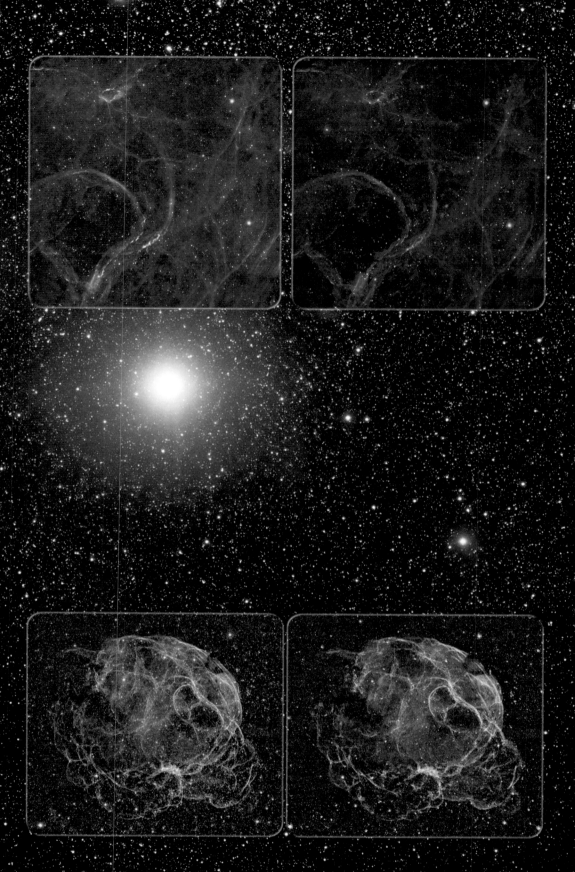

The Spaghetti Nebula

One of the sky's most fascinating supernova remnants is Simeis 147 (Sharpless 2–240, and sometimes called the Spaghetti Nebula). Lying on the border of the constellations Auriga and Taurus, this very faint nebula is some 3,000 light-years away and is the product of a star that exploded some 40,000 years ago. The bright star is Elnath, Beta Tauri.
Credit: Rogelio Bernal Andreo

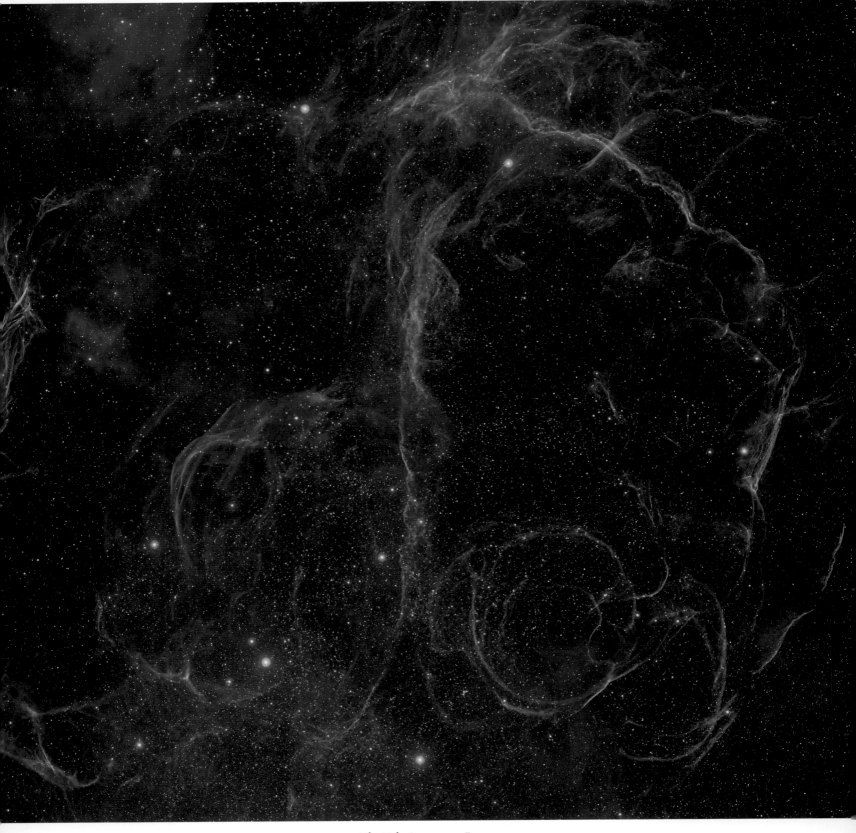

The Vela Supernova Remnant

The Vela Supernova Remnant lies in the deep southern sky at a distance of some 800 light-years, and is the tangled mess of debris from a massive star that exploded some 8,000 years ago. In 1968 astronomers studied a pulsar lying nearby, the Vela Pulsar, which is a rapidly rotating star composed nearly entirely of neutrons. From this study, they determined that these exploding stars, supernovae, create neutron stars. *Credit: Don Goldman*

The sky offers many more examples of supernova remnants, reminders of the "live fast and die young" example set by the most massive stars we know about. Observing these magnificent objects, whether it be through a telescope

The Puppis A Supernova Remnant

The widely dispersed supernova remnant known as Puppis A lies in the southern sky and stretches over a diameter of about one degree, twice as wide as the Full Moon. The debris from a massive star that exploded some 3,700 years ago, this gas cloud partially overlaps the Vela Supernova Remnant in our sky, but the two objects are unrelated, the former lying four times farther away, at a distance of about 6,500 light-years. *Credit: Don Goldman*

or gazing at beautiful pictures in a book, reminds us of the incredible complexity and size of the cosmos, and of the finite life spans of the stars.

The Ring Nebula

One of the most famous planetary nebulae in the sky, the Ring Nebula (M57) in Lyra is easily visible in small telescopes as a smoke-ring shaped glow. It lies some 2,600 light-years away, and spans about 1.3 light-years across. Its 15th-magnitude central star can be glimpsed in very large backyard telescopes.
Credit: J.-P. Metsävainio

The Future of the Sun, and Us

We know that the Sun will exhaust its nuclear fuel about five billion years from now, swell into a red giant, and undergo the process that results in a planetary nebula, eventually leaving just a white dwarf remnant when the nebula has dissipated into the galaxy. Of the several possible shapes of a planetary, from a spherical shell to a flattened, bipolar nebula, do we know what our solar system will look like? No. But we can glean a good survey of the possibilities by examining some of the most spectacular planetary nebulae in the sky, most of which are visible in modest backyard telescopes.

Perhaps the most observed planetary nebula in the sky is the Ring Nebula (M57) in Lyra, which shows a classic, oval "smoke ring" shape magnificently in small telescopes. It is easy to find, being lodged between the bright stars Beta and Gamma Lyrae, and can be seen in a three-inch telescope. The nebula is relatively small but it is bright, and has what astronomers call a high surface brightness. That is, its light is not spread out so much as to appear faint. The nebula appears like an oval, gray-green glowing ring, with a prominent central hole. (Photographs capture a broad range of color in astronomical objects. Our human eyes do not build up light and our faint light receptors are not sensitive to color. Therefore, faint astronomical objects appear to us, as viewed in telescopes, as mostly a shade of gray-green.)

The Ring Nebula may give us a good representation of what our future solar system will look like. It is some 2,600 light-years away, and spans about the size of our solar system, some two light-years across. The dim central star glows faintly at 15th magnitude (see Glossary), making it visible only in quite large backyard telescopes.

In the northern hemisphere, the Ring Nebula is well placed to be observed in the summertime evening sky. And other planetaries are visible in the same sky, at the same time. Another terrific example is the Dumbbell Nebula (M27) in Vulpecula (see chapter 3).

One of the closest bright planetary nebulae to us is the Helix Nebula (NGC 7293), which lies in the autumn evening sky in the constellation Aquarius. A mere 655 light-years away, this object has a complex, spiraling shape that almost resembles the double helix of a DNA molecule, thus the nickname. The Helix is large enough in our sky – a good fraction as large as the Full Moon – but it has a relatively low surface brightness, and so it is a bit more challenging to see in a backyard telescope.

Close-up images of the Helix Nebula made with the Hubble Space Telescope show a multitude of unusual features. This was the first planetary nebula seen to contain cometary knots, dusty globules that are seen with tails, material that is blowing away from them as the stellar wind that creates the nebula is moving outward. Astronomers believe the Helix is about 10,000 years old.

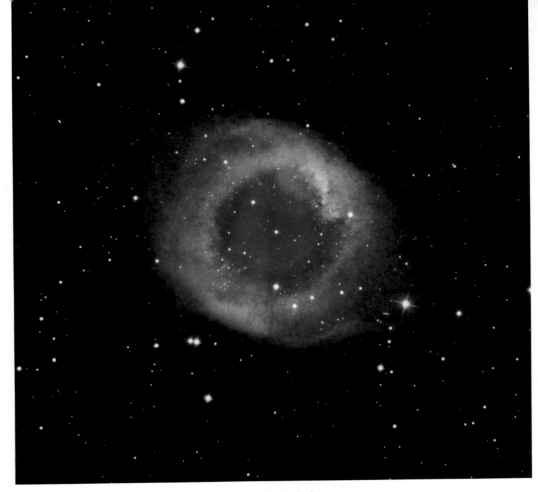

The Helix Nebula

The spectacular Helix Nebula (NGC 7293) in Aquarius is one of the closest planetary nebulae to us, lying just 655 light-years away. The curling form of the gas reminds us of the DNA double helix, thus leading to its distinctive name. The central star that gave rise to the nebula is plainly visible in backyard telescopes, and astronomers believe the nebula formed about 10,000 years ago. *Credit: J.-P. Metsävainio*

The springtime evening sky in the northern hemisphere hosts a beautiful planetary in the form of the Owl Nebula (M97), located near the bowl of the Big Dipper (the Plough) asterism in the bright constellation Ursa Major. This glowing, circular disk of light is relatively small and contains two dark "eyes" on either side of the dim central star, thus giving rise to its name. A challenging object to observe, the Owl appears somewhat dim to most observers, and it is comfortably viewed in a six-inch or larger telescope. Discovered by French astronomer Pierre Méchain in 1781, the nebula contains three concentric shells, the "eyes"

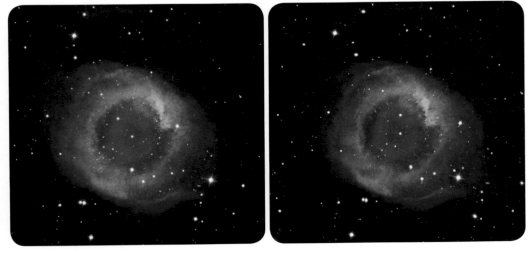

The Helix Nebula in 3-D

Often referred to as the "Eye of God," the impact of this fabulous nebula is heightened when seen in three dimensions.

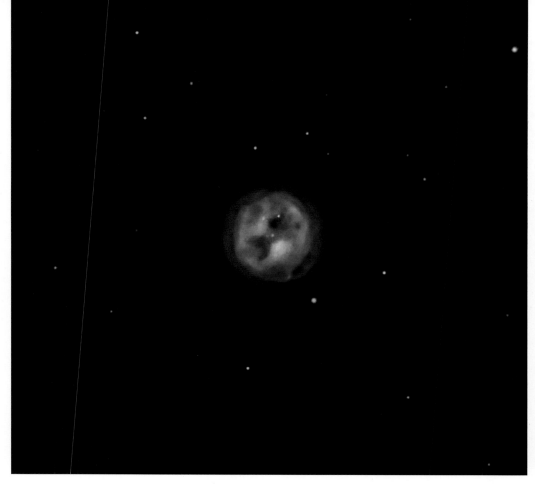

The Owl Nebula

The Owl Nebula (M97) in Ursa Major is one of the sky's most alluring planetary nebulae, with two dark "eyes" that seem to stare at viewers. The remains of a Sun-like star that extinguished its fuel about 8,000 years ago, the nebula lies some 2,000 light-years away and contains several concentric shells of extremely faint gas. *Credit: J.-P. Metsävainio*

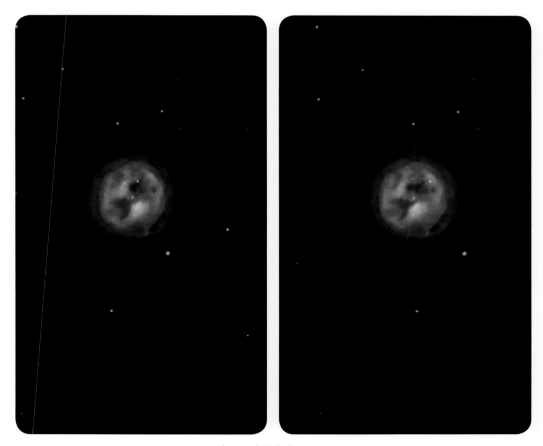

The Owl Nebula in 3-D

The eyes of the Owl become part of a glowing sphere with dusty patches when viewed in 3-D. The faint central star is visible, and its companion sun fades far into the background.

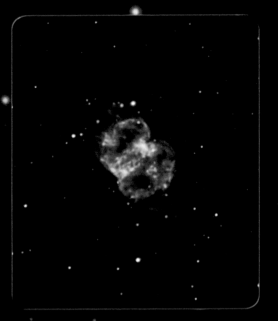

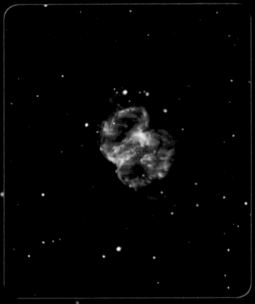

The Little Dumbbell Nebula

One of the brightest planetary nebulae in the sky is the Little Dumbbell Nebula (M76) in Perseus. The cloud has a twin-lobed appearance vaguely reminiscent of the larger Dumbbell Nebula, and lies some 2,500 light-years away. About the same size as our solar system, the Little Dumbbell is a popular target for observers on winter nights.
Credit: Adam Block/Mount Lemmon SkyCenter/University of Arizona

formed by the fact that an inner shell is not symmetrical, but is an off-axis barrel-shaped structure. The faint central star, at 14th magnitude, can nonetheless be spotted in large amateur scopes. This nebula lies some 2,000 light-years away.

In the wintertime evening sky, observers can spot the Little Dumbbell Nebula (M76) in the bright constellation Perseus. Named for its resemblance to M27 (see chapter 3), the Little Dumbbell has a bright "bar" structure and two lobes that emanate from its ends, giving it a slightly differently structured dumbbell shape than its brighter cousin in Vulpecula. Nonetheless, this object is an attractive target for backyard observers and it can be seen in a four-inch scope. Discovered by Méchain in 1780, it is a bipolar nebula lying some 2,500 light-years away. Its central star glows dimly at 16th magnitude, making it a challenge to see even with large backyard telescopes.

The southern hemisphere sky also contains a retinue of unusual and attractive planetary nebulae. One of the most distinctive is the Bug Nebula (NGC 6302), a bipolar nebula that appears to be somewhat distorted, lying in the constellation Scorpius. This object's central star is extraordinarily hot, burning at 450,000 degrees, and the star that gave rise to the nebula must have been on the massive end of the range that will produce a planetary nebula. A dense belt of dust encircles the equator of this star, and has helped to constrain the outflow of gas and make this nebula so focused into two lobes. This strange object lies at a distance of 3,400 light-years.

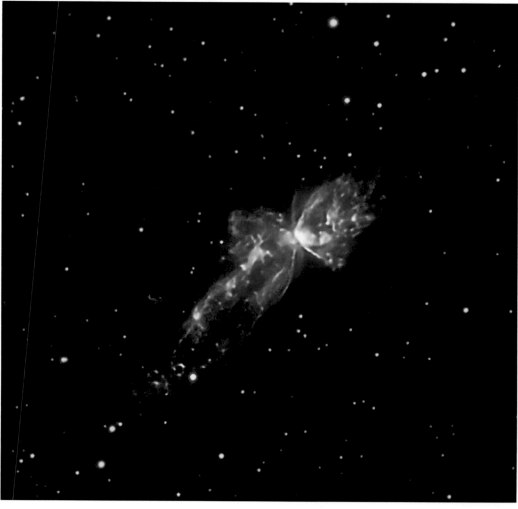

The Bug Nebula

The Bug Nebula (NGC 6302) in Scorpius is a distinctive bipolar planetary nebula with a flattened shape, the gas squeezing outward from the star's poles. Located 3,400 light-years away, it is a fine challenge for observers with moderate-size telescopes. *Credit: J.-P. Metsävainio*

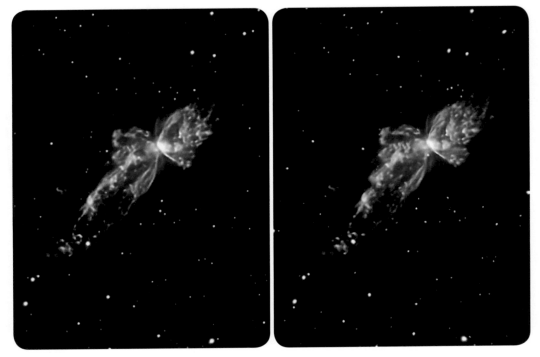

The Bug Nebula in 3-D

NGC 6302 takes on a dramatically new shape in stereo, with its broad, faint body in the backdrop, and the brighter "wings" jumping out at viewers. The strongly bipolar shape comes from dust preventing the gas from escaping except along the two axes.

Lying in the southern constellation Hydra, the Ghost of Jupiter (NGC 3242) is a prominent disk-shaped planetary with a bright oval ring surrounded by a faint outer halo of light. It appears to be about the same angular size as the king of planets, thus creating its name. Located some 1,400 light-years away, this nebula is young, its inner regions having expanded for some 1,500 years. Another southern nebula, this one in Vela, is the Eight-Burst Nebula (NGC 3132), a bright, ring-shaped cloud of gas that lies 2,000 light-years away. A bright ridge of gas surrounds the faint central star, and a hazy cloud of fainter gas surrounds the whole assemblage in this unusual object.

The Ghost of Jupiter Nebula

This ghostly image from NASA's Spitzer Space Telescope shows the disembodied remains of a dying star. The Ghost of Jupiter (NGC 3242) lies roughly 1,400 light-years away in the constellation Hydra. This infrared view shows off the cooler outer halo of the star, colored red. Also on view are concentric rings around the object, the result of material being periodically thrown off in the star's final death throes. *Credit: NASA/JPL-Caltech/Harvard-Smithsonian CfA*

The Eight-Burst Nebula

NGC 3132 is a bright planetary nebula lying in the southern constellation Vela. This object's complex structure contains a bright, inner shell that glows in blue-green hues surrounded by older gas expanding into a chaotic pattern around it. The nebula lies some 2,000 light-years away. *Credit: Adam Block/Mount Lemmon SkyCenter/University of Arizona*

NGC 2899

Lying deep in the southern sky, planetary nebula NGC 2899 in Vela was discovered by John Herschel in 1835. Lying at a distance of some 6,500 light-years, this nebula shows a pronounced elongated shape, and appears to be a bipolar object. *Credit: Don Goldman*

The Robin's Egg Nebula

The bright blue-green planetary nebula NGC 1360 is sometimes called the Robin's Egg Nebula. In the southern constellation Fornax, it glows faintly but is strongly visible in the oxygen III bands of light, giving it its hue. Lying at a distance of about 1,800 light-years, it is the remains of the dying star, visible in its center, that exhausted its nuclear fuel several thousands of years ago. *Credit: Adam Block/Mount Lemmon SkyCenter/University of Arizona*

The sky is filled with many other unusual examples of planetaries that show a wide range of shapes, and you can enjoy exploring them with a telescope or simply in beautiful astroimages from your armchair. NGC 1360 in Fornax, sometimes called the Robin's Egg Nebula, shows an oval disk of light that is strongly blue in color due to strong oxygen emission. In Eridanus, the tiny, circular nebula NGC 1535 is sometimes referred to as Cleopatra's Eye. This curious object displays a smooth envelope of nebulosity but adds some brighter, mottled lines and twists of gas within the inner sanctum.

Back in the northern sky there are many little-known treasures, too. One of the most beautiful spherical nebulae in the summer Milky Way is NGC 6781 in

NGC 1535

Discovered by William Herschel in 1785, the bright blue-greenish planetary nebula NGC 1535 in Eridanus lies at a distance of about 6,000 light-years. Some observers refer to it as the Cleopatra's Eye Nebula. *Credit: Adam Block/Mount Lemmon SkyCenter/University of Arizona*

Aquila, often overlooked because of the more well-known Ring and Dumbbell nebulae lying not far away. The bright, cross-shaped constellation Cygnus contains a small hoard of interesting planetaries. They include NGC 6826, called the Blinking Planetary because it helps to show off an artifact of astronomical observing as well as any object in the sky. When we look directly at an object, we're employing the cones, the bright light receptors in the center of our eyes. Glancing slightly to the side enables our rods, our faint light receptors. Centering NGC 6826 in a telescope's eyepiece, we see its bright central star when staring at the nebula, and then the faint nebula pops into view when glancing to the side of the field. Moving one's eye back and forth makes the nebula "blink," thus the name.

NGC 6781

One of the finest planetary nebulae visible in the autumn evening sky is NGC 6781 in Aquila. Its smooth, ring-shaped shell of gas stands out in a rich field of Milky Way stars. It lies about 3,000 light-years away. *Credit: Adam Block*

NGC 7027

The young, dense planetary nebula NGC 7027 in Cygnus features a boxy shape with an inner core, a faint outer halo, and high-velocity gas that is emerging through the edges on several sides. It lies about 3,000 light-years away.
Credit: Adam Block/Mount Lemmon SkyCenter/University of Arizona

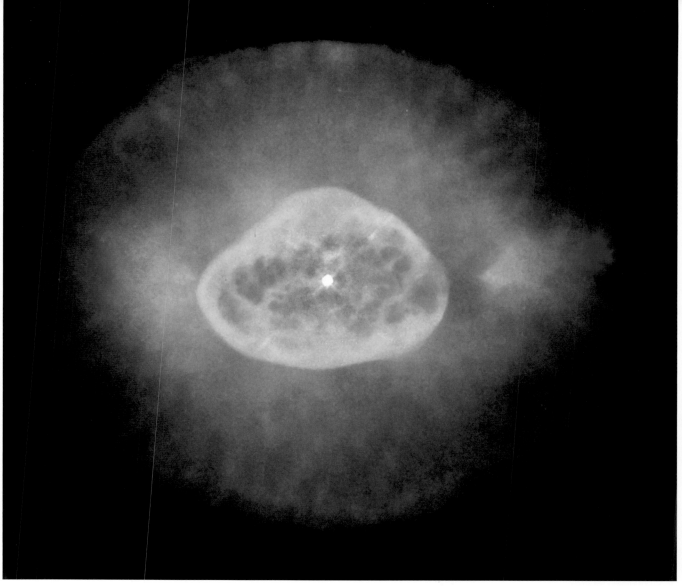

The Blinking Planetary

NGC 6826's eye-like appearance is flanked by two red patches in this view from the Hubble Space Telescope. The faint green "white" of the eye is believed to be gas that made up almost half of the star's mass for most of its life. The hot remnant star (in the center of the green oval) drives a fast wind into older material, forming a hot interior bubble which pushes the older gas ahead of it to form a bright rim. *Credit: Bruce Balick (University of Washington), Jason Alexander (University of Washington), Arsen Hajian (U.S. Naval Observatory), Yervant Terzian (Cornell University), Mario Perinotto (University of Florence, Italy), Patrizio Patriarchi (Arcetri Observatory, Italy) and NASA/ESA*

Cygnus also holds treasures in NGC 7027, a boxy-shaped nebula that is a young planetary, perhaps only several hundred years old. It also holds a highly-evolved and strange planetary in NGC 7008, sometimes called the Fetus Nebula. This object's speckled face, lying in a rich star field, makes it an attractive target. And dozens of other planetary nebulae lie scattered all across the sky, foreshadowing what our little planetary neighborhood will look like long after we are no more.

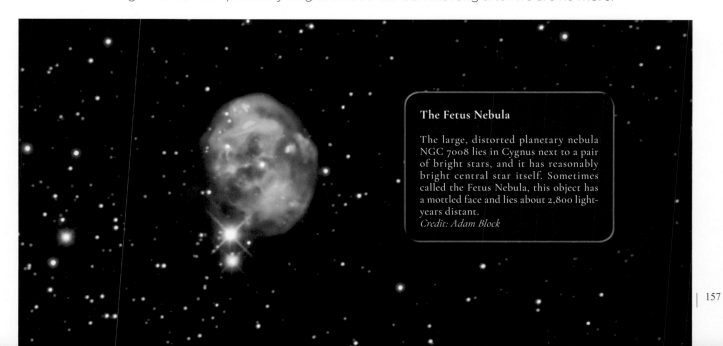

The Fetus Nebula

The large, distorted planetary nebula NGC 7008 lies in Cygnus next to a pair of bright stars, and it has reasonably bright central star itself. Sometimes called the Fetus Nebula, this object has a mottled face and lies about 2,800 light-years distant.
Credit: Adam Block

THE END OF THE WORLD?

Whatever form our Sun's planetary nebula will take, the solar system will transform dramatically long before that ever happens. The Sun is actually a variable star, not in the sense that astronomy enthusiasts usually think of, with regular up-and-down variations in light output that can be observed from great distances. But it is slowly increasing its total energy output over time, and this will have huge effects on our planet over the coming few hundred million years. Of all the crazy things that change on our planet, in our lives, every day, we tend to think of the Sun as one great constant that is always there, in the same way, every time we look up. But it's not a constant.

There's no reason to panic for humanity. Creatures like us have only been around for about two million years, although life on Earth dates back to at least 3.5 billion years ago. But over the next 1.5 billion years, Earth's global surface temperature will rise to 80° C (176° F), approaching the boiling point of water. That's independent of human-induced climate change, and with more carbon dioxide in the atmosphere, that point could come sooner. Studies show that Earth's oceans could be gone in a billion years or less. The last water molecules on our planet could be gone within 2.5 billion years. That will mark the endgame for life on Earth. But for the most fragile life on the planet, including humans, the end will come far sooner than that.

So, it's clear that life on our planet will end far sooner than the four to five billion years from now when our Sun will become a red giant and then produce a planetary nebula. Visions of that spectacle will have to be made by other civilizations in other parts of the galaxy, long after Earth's civilization is no more. It is somewhat astonishing, given that the earliest life on our planet is 3.5 billion years old, and with about 1.5 billion years of life left, that about three-quarters of the history of life on Earth has already taken place.

When Earth's water is gone, no more carbonate rock weathering will take place, a process that helps to store away vast quantities of carbon dioxide. That will lead to a runaway increase of carbon dioxide, transforming Earth into a Venus-like world. Our "blue marble" will become an unrecognizably hellish world that will no longer support life.

Can human civilization survive that awful transformation, assuming that it still exists by then? Some scientists have suggested terraforming planets, creating hospitable worlds from unhospitable ones. Could Mars be made into a livable world in which large numbers of people could actually be there for long periods? Early studies suggest that reengineering the levels of carbon dioxide there could raise the temperature above the freezing mark, and that covering the polar caps with dark material could release substantial amounts of carbon dioxide into the atmosphere. If large amounts of volatiles exist under the martian surface as they should, perhaps they could also be released, in order to raise the planet's temperature to ranges that would allow humans to live there.

But in realistic terms, trying to reengineer a planet's atmosphere is a far larger engineering project than anything ever undertaken by humans thus far. The effort that would be required would be beyond anything ever attempted by our civilization, and many scientists look at such questions as hypotheticals to understand planetary science and how it works, but not something that could actually be done in reality. Moreover, if we're looking at jumping out to the next terrestrial planet to stave off the end of life on Earth, transforming Mars would only help for a brief time before the whole solar system would be uninhabitable. With the Sun's increased radiative output, the whole solar system in reality will be a bad place for living things to be.

Let's also consider this question of how dangerous a place the universe really is. Again, we don't think about this stuff on a day-to-day basis. We live our lives on Earth's surface, walking around what seems like a two-dimensional plain. Rarely do most people look skyward and remember that an enormous galaxy full of stars is out there, can affect us, and that we're part of it during every second of time, whether we're contemplating it or not.

DIAMETER OF OUR SUN AFTER TRANSITIONING TO A RED GIANT

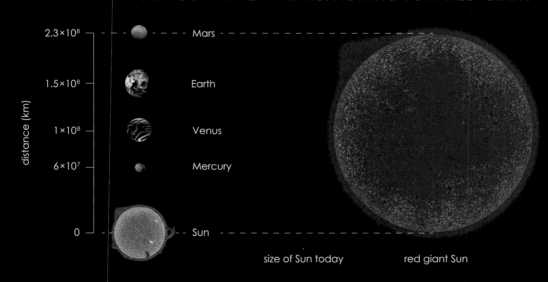

Mass Extinction Events

We may not like this idea, but the universe could step in and stop the grand human experiment at any time. To appreciate the possibilities, we should briefly think of the few extinction events in history that we know a fair amount about. The most famous mass extinction event is the Cretaceous-Paleogene (also known as K-Pg – from the German "Kreide" (chalk), and Pg "Paleogene") Extinction Event, better known as "The Event." This happened 66 million years ago when an asteroid came flying into what is now the Yucatán Peninsula in Mexico, and which, among other things, killed the dinosaurs, probably making our species' rise to significance possible at all. The K-Pg event killed 17 percent of all living families, 50 percent of all genera, and 75 percent of all species on Earth. In the age of renewed life, it enabled mammals and birds to emerge as dominant.

But the causes of mass extinctions are far more varied than a rock hurtling into our planet from space. Beside asteroid and comet impacts are other dangers – changes in sea level, global warming, global cooling, marine anoxia (lack of oxygen for sea creatures), changing ocean-atmosphere circulation patterns, increasing solar radiation, plate tectonics, and Large Igneous Province volcanism (eruptions from special geological hotspots). These are real, natural dangers, independent of human-induced dangers that we are piling on top of the huge risks for our planet, such as thoughtlessly spewing far more carbon dioxide into the atmosphere than we might, or filling the oceans with life-choking plastics.

If human leaders were smart, they would learn from the five major extinction events we know about, one of which was the K-Pg. Others include the Triassic-Jurassic extinction event, some 201 million years ago. This one killed 23 percent of all families, 48 percent of all genera, and 75 percent of all species. It paved the way for the dominance of the dinosaurs by eliminating many of their competitors.

The Permian-Triassic extinction event took place some 252 million years ago. It is the largest mass extinction known in our planet's history, killing 57 percent of all families, 83 percent of all genera, and perhaps 96 percent of all species. This was the so-called "great dying" event that fundamentally changed the course of life on Earth.

The Late Devonian extinction event, about 375 million years ago, consisted of a series of extinctions that reduced the number of families by 19 percent, the number of genera by 50 percent, and the number of species by 70 percent. This series of events may have lasted for some 20 million years.

And the Ordovician-Silurian extinction event occurred about 444 million years ago and consisted of two events that killed 27 percent of all families, 57 percent of all genera, and as much as 70 percent of all species.

Life on our planet is a pretty fragile thing, and it has experienced numerous traumatic events that have changed its course over history. There were undoubtedly many others farther back in time. And mere knowledge does not protect us from reality: because we know about these now doesn't mean that somehow they'll magically stop.

Dangers that we know of from space will have to be contested with. Some 20,000 Near-Earth asteroids and comets have been discovered. There are no "civilization killers" of the K-Pg Impact size in near-Earth space, but that doesn't mean one couldn't come from farther out that would impact our planet and wipe us out. More likely, a smaller object could be huge trouble, if not wipe out our whole civilization. We know about only one percent of the asteroids of one kilometer diameter in the inner solar system, and these objects could be devastating if one struck us, which over long enough intervals, will certainly happen again.

And more distant, more sinister effects could spell enormous trouble for our civilization. What about a nearby supernova exploding and bombarding our star system with sterilizing radiation? A gamma-ray burst, another type of explosive event, could also do this. Antimatter encountering our solar system, or a passing mini black hole, could spell doom. In those cases, we would never know what hit us.

Although we may not look upward into the sky as frequently as we should, the universe has lessons for us. All we have to do is pay attention. The main lesson is that life is short. Individual lives and even civilizations are but the blink of an eye in the cosmic timetable of the heavens. We need to take care of our planet rather than selfishly trashing it and abusing its resources without giving a thought toward future generations. We need to take care of each other, and all of our fellow creatures too, who have just as much right to live comfortable and enriching lives as we do. We not only share the planet with other species, but we depend on their success just as much as they do.

When will humans learn the lessons that will help us do the right things for the sake of our planet and for each other? Perhaps the answers lie in the gazes we do make skyward, to glance at the Sun that powers our world, or even, in a cool, dark night, as we peek through a telescope at one of those cosmic clouds. These distant nebulae remind us of where we came from, where we are going, and that we are all together in this grand experiment of the cosmos.

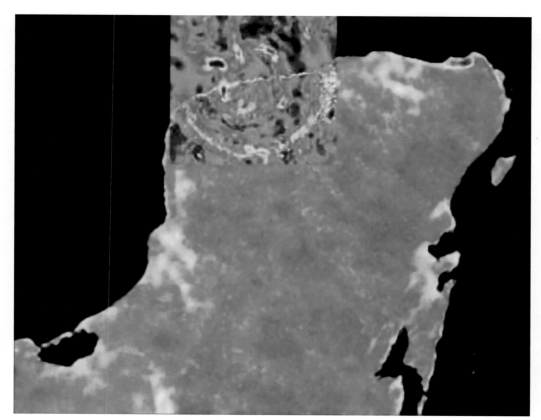

The Chicxulub Crater

The Chicxulub Crater is the result of an asteroid impact that wiped out 75 percent of all living species on Earth. It now lies mostly underwater just off the Yucatán Peninsula in Mexico. From rim to rim it measures some 180 kilometers (112 miles) across and nearly 20 kilometers (12.5 miles) deep. NASA's gravity anomaly map shows gravity highs in yellow and red, and gravity lows in green and blue. It is superimposed on a map of the peninsula. *Credit: NASA*

The Authors at Meteor Crater

The authors on a visit to Meteor Crater (Barringer Meteorite Crater) in Arizona soon after the launch of their first collaborative publication *Mission Moon 3-D* in 2018. The crater was formed around 50,000 years ago when an asteroid 46 meters (150 feet) across plunged to Earth. The crater is 1,000 meters (3,900 feet) in diameter and about 170 meters (560 feet) deep.

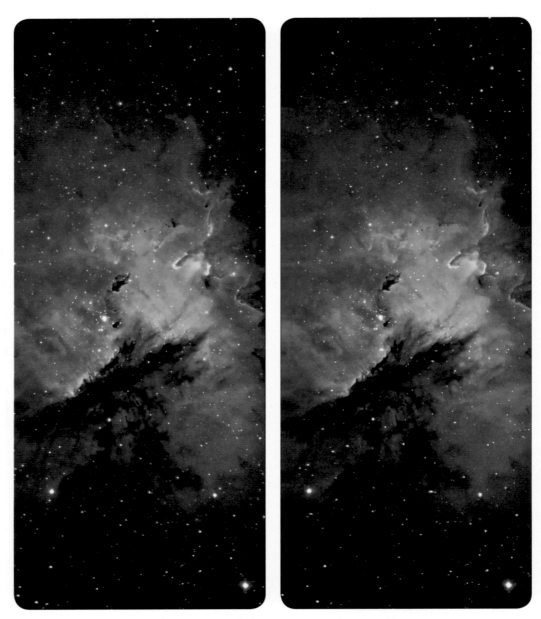

The "Pacman Nebula" in Cassiopeia (NGC 281)

Note how the dark nebula floats in front of the ionized regions of gas. The open cluster IC 1590 at the center of the nebula, just behind the dark cloud of gas, is the source of radiation and solar wind that ionized and shaped the nebula. There are several dark Bok globules floating inside the nebula. The pillar-like structures are formed from the collapsing gas and they all are pointing to the source of the stellar wind, the open cluster IC 1590 at center. The bluish tone is from ionized oxygen, O-III, and can only be seen around the star cluster at center, since it needs lots of energy to ionize. Hollow spaces inside the nebula form when the solar wind is blowing the gas away around IC 1590.

How the Stereos Were Made

J.-P. Metsävainio

I have worked as an astronomical photographer and artist in northern Finland for over 20 years. Because I'm located not too far from the Arctic Circle, winters are very dark up here, but unfortunately also very cloudy. At this latitude we have about six months during which it is not dark enough for deep-sky astrophotography. Having so few nights suitable for astronomical imaging has given me more time to think, and it also forced me to develop some new astronomical imaging and processing techniques. One of them is a method to convert distant gas clouds to the 3-D-models.

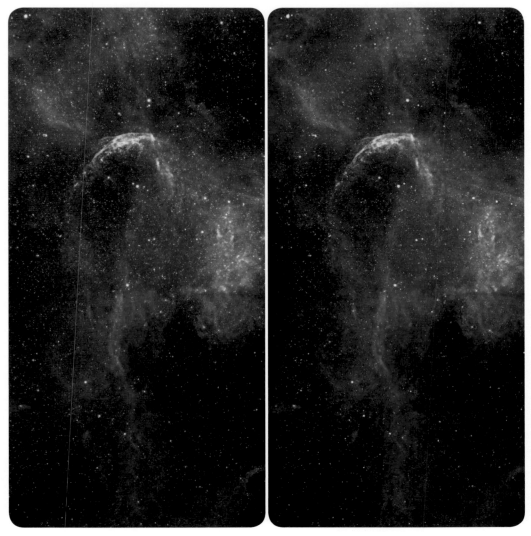

A Bubble around the Wolf-Rayet star (WR 134) in Cygnus

The star WR 134 is a variable Wolf-Rayet star in Cygnus. It is the white star just above center of the image, inside the faint ring structure. This aggressively burning hot star is in a final phase of massive star evolution to a spectacular supernova explosion. A strong stellar wind from the hot massive star forms a bubble-like shell around the star. The surrounding shell is over 130 light-years wide. The blue arch is ionized oxygen, O-III.

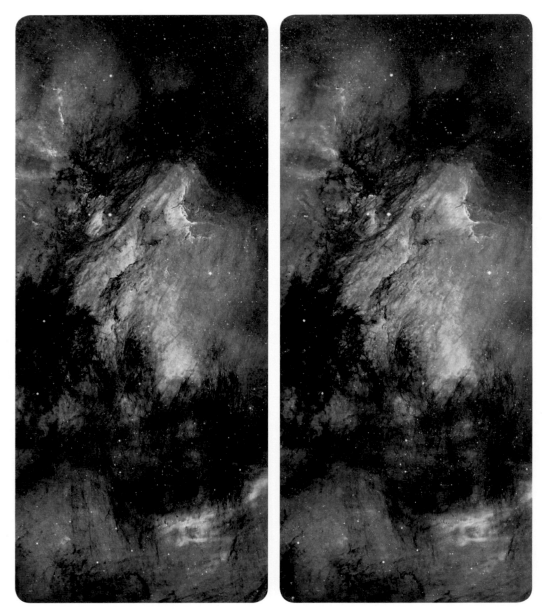

A Bridge of Dark Gas between the North America and Pelican nebulae

This region of bright emission that comprises the North America and Pelican nebulae is very well known by astronomical imagers around the world. I made this 3-D model to show that those two separate emission areas are actually just one big region of gas formation. It is divided visually by a bridge of dark unionized gas in front of it.

If you look at a picture of a nebula, it looks like a painting on a canvas, totally flat. In reality nebulae are naturally three-dimensional objects floating in three-dimensional space. Since celestial distances are so vast, it's not possible to obtain two images from slightly different angles to form a true stereoscopic image and see the actual shapes of the gas cloud.

For as long as I have captured images of celestial objects, I have always seen them three-dimensionally in my head. Over time I realized that we actually have enough scientific information to build a coarse skeleton model of the nebula itself. The scientific information makes my visions much more accurate, and the 3-D technique I have developed enables me to share those beautiful visions with others.

How accurate my 3-D-visions are depends on how much accurate information I have and how well I implement it. Also, many different estimates are needed for the 3-D model. The final 3-D-image is always an appraised simulation of reality based on known scientific facts, deduction, and some artistic creativity on top of everything else.

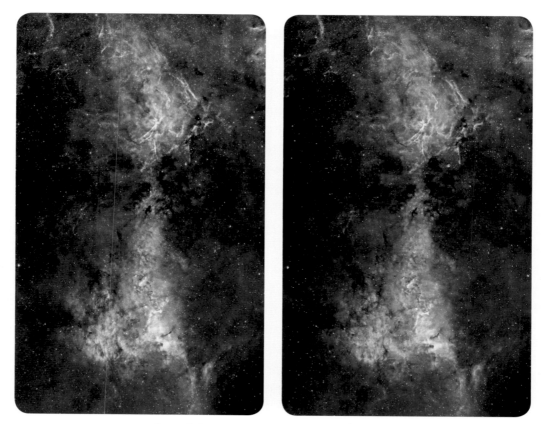

Central Cygnus: LBN 251, LBN 239 and IC 1311

I selected this photo as an example of how 3-D vision can clarify things. This complex area in central Cygnus shows hot young stars ionizing the gas to make it glow, while at the same time radiation pressure sweeps the gas about and forms a hollow space around the star cluster. At close range, the radiation has enough energy to ionize oxygen, which shines in bluish hues. There are complex filaments of unionized gas floating in front of the glowing areas.

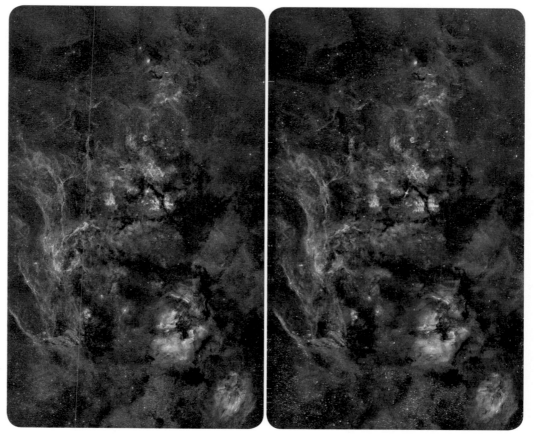

An 18-panel Mosaic of the Constellation Cygnus

This stereo pair shows almost the entire constellation of Cygnus. In the lower right-hand corner are the North America and Pelican nebulae. A small, bright area near the center is the Tulip Nebula. A small blue region above the center of the image is the Crescent Nebula. The faint blue ring right of center is the supernova remnant W63. Complex filaments of unionized gas are floating in front of the emission areas.

After I have collected all the necessary scientific information about my target, I start my 3-D conversion using the stars in the image. Usually there is a recognizable star cluster which is responsible for ionizing the nebula. We don't need to know its absolute location since we know its relative location. Stars ionizing the nebula have to be very close to the nebula structure itself. I usually divide up the rest of the stars by their apparent brightness, which can then be used as an indicator of their distances, brighter being closer. If true star distances are available I use them, but most of the time my rule of thumb is sufficient.

By using a scientific estimate of the distance of the Milky Way object, I can then locate the correct number of stars in front of it and behind it.

Emission nebulae are not lit up directly by starlight; they are usually way too large for that. Rather, stellar radiation ionizes elements within the gas cloud. So it is the nebula itself that is glowing, at the characteristic wavelengths of each ionized element. (The principle is very much the same as in fluorescent tubes.) I use this information for my 3-D model. The thickness of the nebula can be estimated from its brightness, since the whole volume of gas is glowing, brighter means thicker. By this means, forms of the nebula can be turned to a real 3-D shape. Nebulae are also more or less transparent, so we can see both sides of it at the same time, and this makes model-making a little easier since not much is hidden.

The local stellar wind, from the star cluster inside the nebula, shapes the nebula by blowing away the gas around the star cluster. The stellar wind usually forms a kind of cavity in the nebulosity. The same stellar wind also initiates the further collapse of the gas cloud and the birth of the second generation of stars in the nebula. The collapsing gas can resist the stellar wind and produces pillar-like formations which must point to a cluster.

Ionized oxygen (O-III) glows with a bluish light, and since oxygen needs a lot of energy to ionize it, this can only be achieved relatively close to the star cluster in the nebula. I use this information to position the O-III area (the bluish glow) at the correct distance relative to the heart of the nebula.

Many other small indicators can be found by carefully studying the image itself. For example, if there is a dark nebula in the image, it must be located in front of the emission nebula, otherwise we can't see it.

Using the known data in this way I build a kind of skeleton model of the nebula. Then the artistic part is mixed with the scientific and logical elements, and after that the rest is very much like creating a sculpture on a cosmic scale.

For more information about how I create stereos and to see my portfolio visit
www.AstroAnarchy.Blogspot.com and www.AstroAnarchy.Zenfolio.com

GALLERY OF STEREO IMAGES

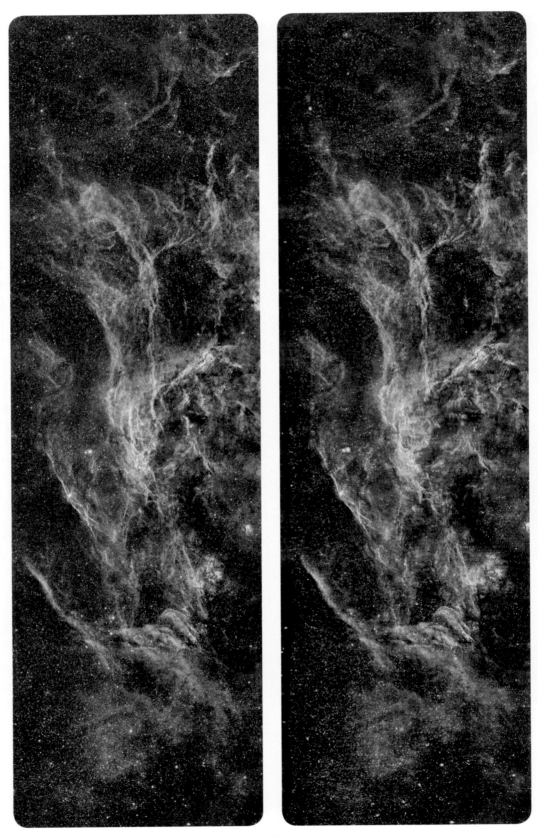

Cirrus of Cygnus

A rare image of cirrus-like gas filaments in the western part of the constellation Cygnus. The filaments are hundreds of light-years long. The Cygnus Shell, supernova remnant W63, is visible at the center of the image, which spans about 15 degrees of sky vertically.

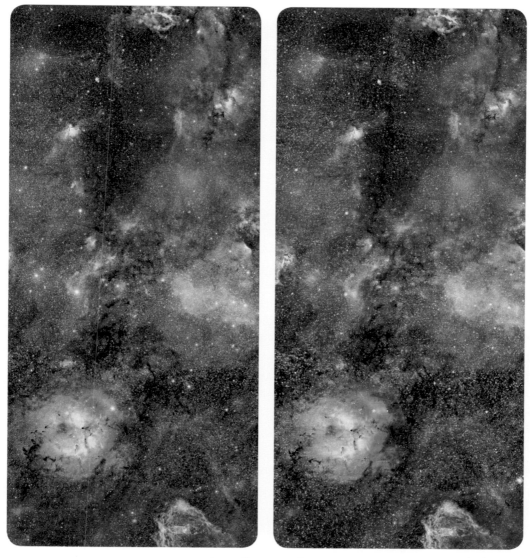

Constellation Cepheus

This 12-panel mosaic shows about 18×10 degrees of the constellation Cepheus. At bottom left lies IC1396, the Wizard Nebula is at upper left and in the upper right corner is the Cave Nebula and Sh2-157. In 3-D the dark gas floats in front of the regions of emission and the correct relative distances between the objects are clearly visible.

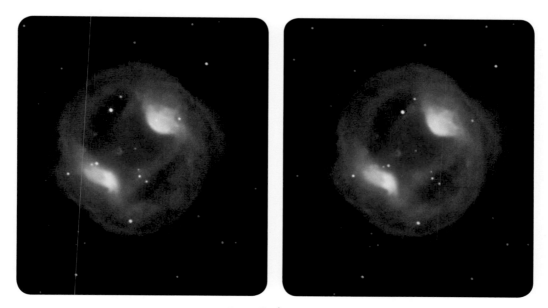

Jones-Emberson 1

Jones Emberson 1 is a large and dim planetary nebula in constellation Lynx. In this 3-D conversion the central star is floating in the heart of the tube-like nebular structure. The outer halo is also visible; the bluish color is ionized oxygen (O-III)

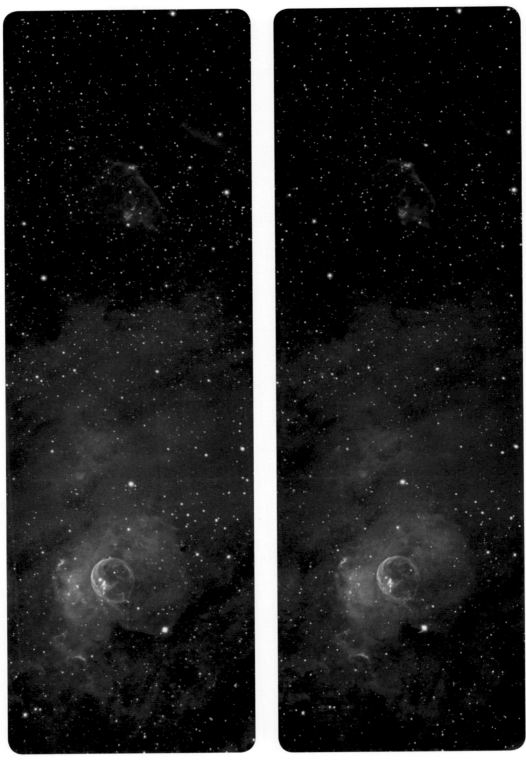

The Bubble Nebula and KjPn 8

This two-panel mosaic shows the famous Bubble Nebula and its rarely seen neighbor the bipolar planetary nebula KjPn 8 (PLN 112-0.1). Relative distances stand out nicely in this 3-D conversion. The Bubble Nebula lies at distance of about 11,000 light-years and the planetary nebula is about 5,500 light-years away.

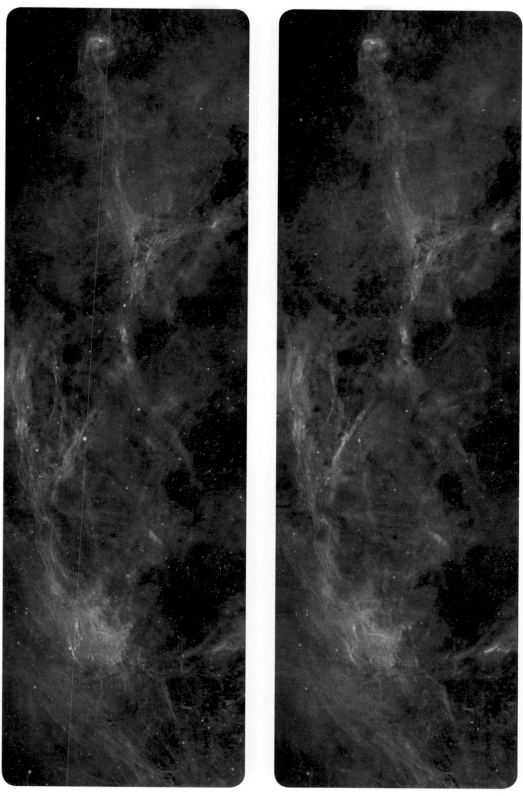

Filaments of Western Cygnus

I took this photo since it shows two different ways to end a star's life cycle. The bluish shape in the lower part of the image is the supernova remnant W63 and the bluish dot in the upper part is the planetary nebula PK 84+9.1

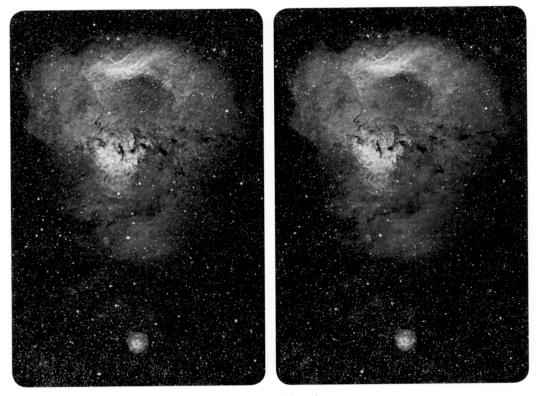

Cederblad 214 and Sharpless 170

I captured this five-degree field in the constellation of Cepheus in 2012, and named it the "Cosmic Question Mark". In this 3-D conversion the distances between the elements are clearly visible. Cederblad 214 lies at a distance of 2,750 light-years and Sharpless 170 at about 7,500 light-years.

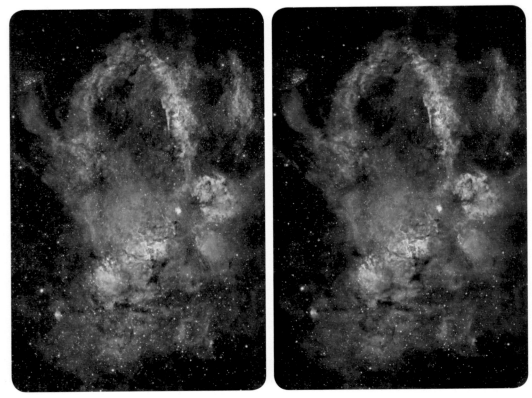

Sharpless 157

For obvious reasons, Sharpless 157 is sometimes referred to as the Lobster Claw Nebula. It is a ring nebula around the Wolf-Rayet star WR 157. The nebulae visible in this image include Sharpless 157, the planetary Nebula WeSb, and the star clusters Markarian 50 and NGC 7510.

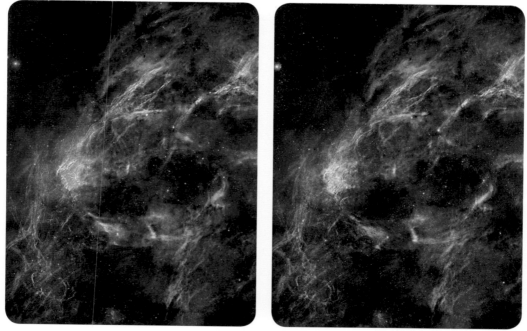

Cygnus shell W63

I first planned to shoot this extremely dim and diffused target many years ago. In autumn 2018 I finally completed my imaging project of supernova remnant W63 in the western part of the constellation Cygnus. The total exposure time was around 200 hours, and it took me five years to collect enough data in the difficult weather conditions in northern Finland.

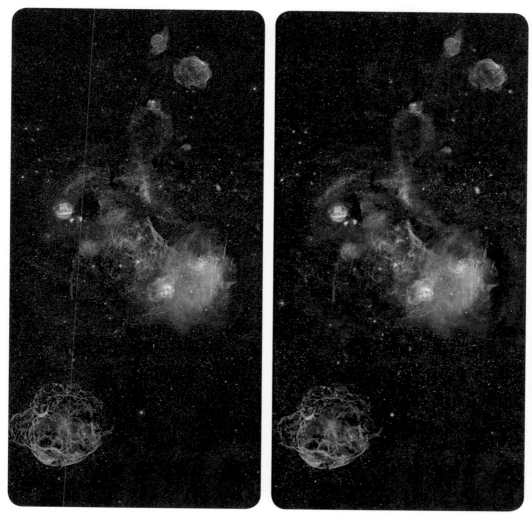

A Mosaic of Auriga

This 12 panel panoramic mosaic spans about 21 degrees of the constellation Auriga. That's 42 Full Moons side-by-side. There are two supernova remnants in this image: Simeis 147 at lower left and Sharpless 224 at upper right. Total exposure time was about 70 hours. Relative distances between different objects are clear to see.

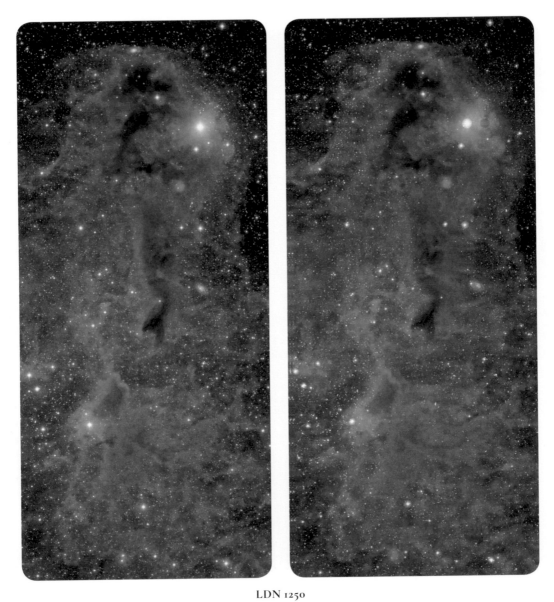

LDN 1250

This image contains objects LDN 1251, LBN 558, PGC 69472, and PGC 166755. This low-mass, star-forming region in Cepheus is an extended cloud of gas and dust.

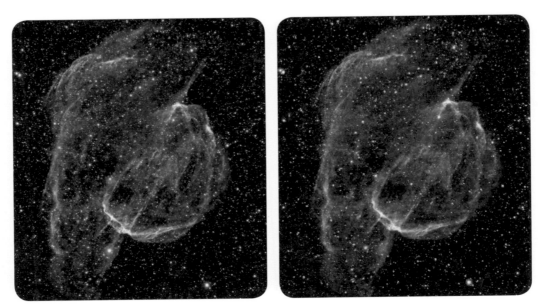

Supernova Remnant in Auriga

Supernova remnant Sharpless 224 is in the constellation Auriga at a distance of about 14,700 light years. It has an apparent size of about 60 arcminutes, and a real size of around 235 light-years. This is a very dim target and lots of exposure time is needed to reveal any detail. Beside hydrogen alpha emission, there are weaker emissions from ionized oxygen, O-III, and ionized sulfur, S-II.

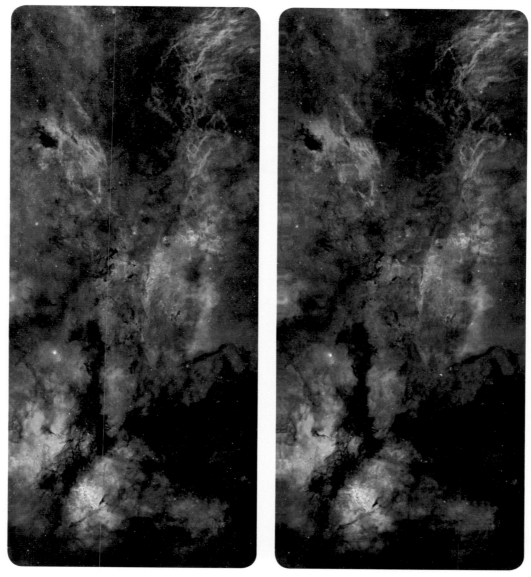

Central Cygnus

In January 2017 I was able to finalize my mosaic image of the central Cygnus region. Eleven photos of gas-filled sky in Cygnus are stitched together seamlessly. The 3-D-conversion reveals how the dark, unionized gas, is shadowing the background emission areas

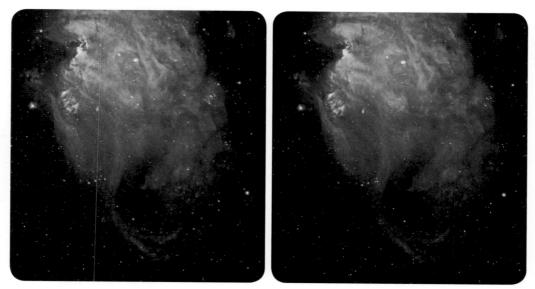

The Monkey Head Nebula

NGC 2174, also known as the Monkey Head Nebula, is located in the Orion constellation at a distance of about 6,400 light years. My photo shows about one square degree of sky (the apparent size of the Full Moon is 0.5 degree). The lower part of the photo shows rarely imaged dimmer parts of the gas formation.

How to Choose a Backyard Telescope

You can view hundreds of nebulae with your own eyes if you have the right kind of telescope, a dark, moonless sky, and allow your eyes to adapt for an hour or so to the darkness.

A telescope collects light. This will allow you to observe much fainter objects than you can see with your eyes alone. The Italian astronomer Galileo described this property in the best way when he said that his telescope "revealed the invisible."

Before choosing which type of telescope to acquire, consider the different types. You may want to find a telescope or camera dealership to look at some models, or attend a star party put on by a local astronomy club. That would allow you to see some live views through a scope before buying.

The most important aspects of a telescope are high-quality optics and a solid, transportable mount. You can buy the best optics that exist, but if they're on a shaky mount, the scope will bring you nothing but frustration. No telescope will work well under windy conditions, but a poor mount will transfer vibrations too easily, so be sure you have a high-quality mount.

Your telescope will require electricity only if you have a motorized drive to compensate for the sky's apparent motion as Earth rotates. If you have a simple "push-pull" mount, you can simply move it yourself, and you won't require electrons.

Several telescope designs exist, and no one is best for everything. If you want to look at faint objects like nebulae, you need as large a mirror or lens as possible, to collect the most faint light, but you don't want a scope that is so large it's hard to use and transport. The best telescope is the one you'll use the most. A smaller telescope that gets used a lot because it's easy to transport is better than a big telescope that mostly collects dust.

The first and simplest type of scope is a refractor, one that uses a lens to collect and focus light at its "sky end," and an eyepiece where the light

Three-inch Reflecting Telescope

With a starter scope like this for about $100/£100 you will be able to see some nebulae, but it is better suited to solar system work such as the Moon, the satellites of Jupiter, and the rings of Saturn.

Five-inch Reflecting Telescope

In addition to magnificent views of solar system bodies, this scope for around
$150/£150 will open up the world of nebulae to the beginning amateur astronomer.

is focused at the other end of a long tube. It's the design that Galileo used. These
instruments generally have a long focus between the lens and eyepiece, and that
makes them "high resolution" instruments best for looking at small, bright objects
like the Moon, planets, and double stars.

Reflectors use a mirror, at the "bottom end" of the scope, and it's easy to
make mirrors, relatively inexpensively, that are much larger than most refractor lenses. So reflecting telescopes are generally easier and more affordable as
options for looking at faint objects like nebulae. A good six-inch or eight-inch
reflector, away from the bright lights of cities, will show numerous nebulae. The
larger the scope's mirror the better, except for that transportation issue. Some
reflectors have simple mounts called Dobsonian mounts that allow you to aim by
rocking them up and down and left to right. They make medium and large size
scopes affordable.

The third major type of telescope is a compound scope, one that uses both mirrors and lenses, and the most popular design of several is the so-called Schmidt-
Cassegrain telescope. These scopes can pack lots of light-gathering power in a
compact, relatively short tube.

Six-inch Schmidt-Cassegrain Telescope

This compact scope (because the light rays are folded back on themselves) for under $1000/£1000 will give the keen amateur astronomer great views of solar system objects and most of the nebulae mentioned in this book.

In recent years, more and more scopes are offered as "go-to" packages, which means they not only have the capability to track the sky's motion, but they have a little onboard computer with a database. That means that once they're aligned, you can simply punch in an object's designation, and the telescope will track around the sky and locate it.

You can learn much more about telescopes from Astronomy.com, or in an issue of *Astronomy* magazine, the world's largest circulation publication on the subject. Telescopes are regularly covered and reviewed in this widely read periodical.

Eight-inch Schmidt-Cassegrain Telescope

Stepping up a level, a scope like this for less than $2000/£2000 with go-to capability (typically loaded with 40,000 objects in the database), is a fabulous platform for taking shots of nebulae.

Dave Eicher and the 24-inch Clark Telescope

Commissioned in 1895 by Lowell Observatory founder Percival Lowell, the firm of Alvan Clark & Sons of Cambridgeport, Massachusetts built the 24-inch (61-centimeter) aperture f/11.3 refractor with a focal length of 22.6 feet (6.8 meters). Since its completion the following year, the telescope has been in regular use to view the heavens and help unravel the wonders of the universe.
Credit: David J. Eicher

Name	Description	Name	Description
Andromeda	Andromeda	Lacerta	Lizard
Antlia	Airpump	Leo	Lion
Apus	Bird of Paradise	Leo Minor	Little lion
Aquarius	Water bearer	Lepus	Hare
Aquila	Eagle	Libra	Balance
Ara	Altar	Lupus	Wolf
Aries	Ram	Lynx	Lynx
Auriga	Charioteer	Lyra	Lyre
Bootes	Herdsman	Mensa	Table
Caelum	Graving tool	Microscopium	Microscope
Camelopardus	Giraffe	Monoceros	Unicorn
Cancer	Crab	Musca	Fly
Canes Venatici	Hunting dogs	Norma	Rule
Canis Major	Great dog	Octans	Octant
Canis Minor	Little dog	Ophiuchus	Serpent
Capricornus	Sea goat	Orion	Orion
Carina	Keel	Pavo	Peacock
Cassiopeia	Cassiopeia	Pegasus	Flying horse
Centaurus	Centaur	Perseus	Perseus
Cepheus	Cepheus	Phoenix	Phoenix
Cetus	Whale	Pictor	Painter
Chamaeleon	Chameleon	Pisces	Fishes
Circinus	Compasses	Piscis Austrinis	Southern fish
Columba	Dove	Puppis	Poop
Coma Berenices	Berenice's hair	Pyxis	Compass
Corona Australis	Southern crown	Reticulum	Net
Corona Borealis	Northern crown	Sagitta	Arrow
Corvus	Crow	Sagittarius	Archer
Crater	Cup	Scorpius	Scorpion
Crux Australis	Southern Cross	Sculptor	Sculptor
Cygnus	Swan	Scutum	Shield
Delphinus	Porpoise	Serpens	Serpent
Dorado	Swordfish	Sextans	Sextant
Draco	Dragon	Taurus	Bull
Equuleus	Foal	Telescopium	Telescope
Eridanus	River	Triangulum	Triangle
Fornax	Furnace	Triangulum Australe	Southern triangle
Gemini	Twins	Tucana	Toucan
Grus	Crane	Ursa Major	Big bear
Hercules	Hercules	Ursa Minor	Little bear
Horologium	Clock	Vela	Sails
Hydra	Sea serpent	Virgo	Virgin
Hydrus	Water snake	Volans	Flying fish
Indus	Indian	Vulpecula	Fox

THE CONSTELLATIONS

The constellations are patterns of stars that appear to lie together in the night sky. There are 88 constellations that are recognized today by the governing body of astronomy, the International Astronomical Union.

Constellation patterns have been drawn up since around BC 4000, but the majority of the names and patterns we are familiar with today come from the ancient Greek, Ptolomy, who rationalized the earlier identifications by the ancient Babylonians, Assyrians, and Egyptians into 48 constellations. He published his list of 48 constellations in the *Almagest*, a mathematical and astronomical treatise on the apparent motions of the stars and planets that was written about AD 150, and remains one of the most important scientific works of all time. In the 16th and 17th centuries many more "modern" constellations were added by European astronomers.

In fact the stars making up each constellation usually lie at vastly different distances from us, since a brighter star at a great distance will appear to have the same brightness as a faint star which is closer.

Overleaf: A 17th century celestial map by the Dutch cartographer Frederik de Wit.
Credit: Public Domain

THE WHOLE SKY MAPS

The night sky on view in the northern and southern hemispheres is shown in these two maps. In addition to showing the patterns for the 88 modern constellations, the brightness of the stars is included, from magnitude 1 to magnitude 5, about the range visible to the naked eye (magnitude 1 being brighter than magnitude 2 and so on; see Glossary). Magnitude 6 stars can be seen in good conditions or with excellent eyesight! In addition, the maps show some of the brighter variable stars, open clusters, nebulae and galaxies.

NORTHERN HEMISPHERE

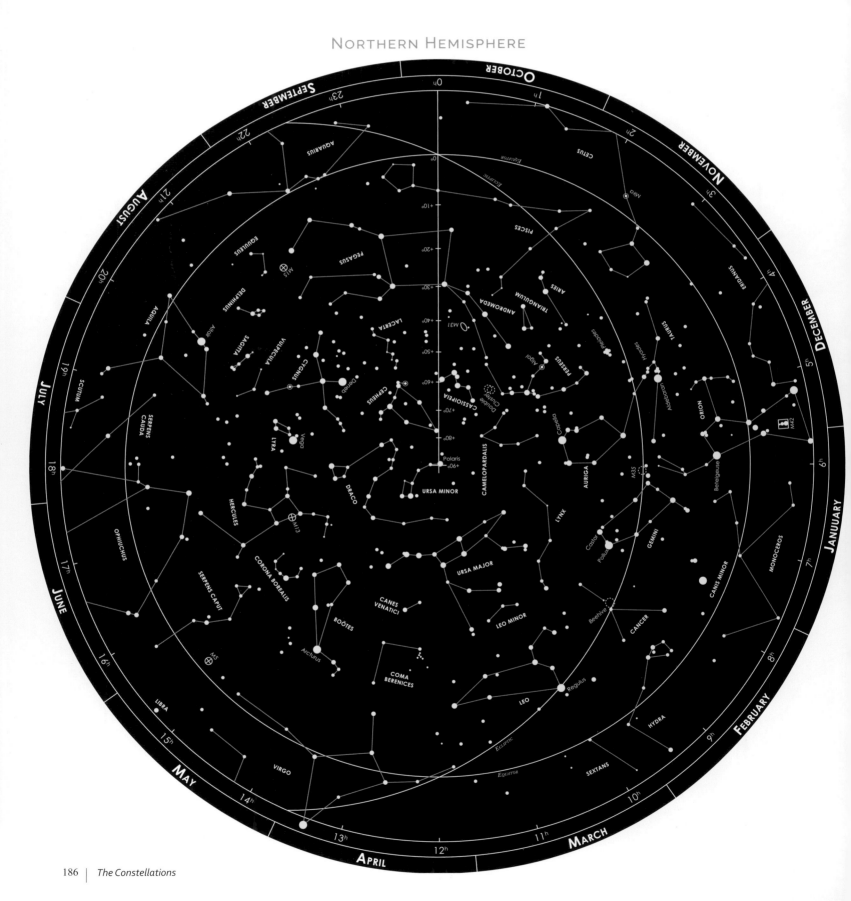

How to Use the Whole Sky Maps

To identify the constellations that you may be able to see in the night sky at any time of year (light pollution and clouds permitting), turn the map for your hemisphere so that the current month is at the bottom. Then the map will show the constellations on view at approximately 2300 h GMT facing south in the northern hemisphere and north in the southern hemisphere. Rotate the map clockwise by 15 degrees for each hour before 2300 h and anticlockwise for each hour after 2300 h.

Southern Hemisphere

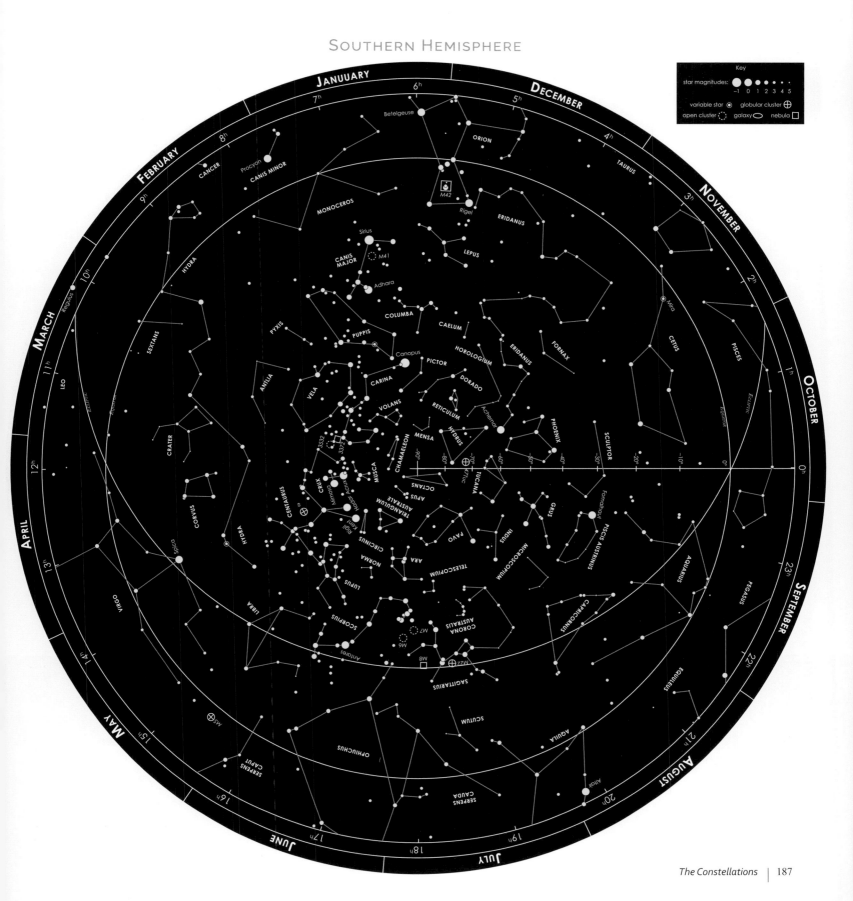

GLOSSARY

ASTRONOMICAL UNIT (AU)
The average distance between Earth and the Sun: 149,598,000 kilometers. Usually rounded up to 150 million kilometers (93 million miles).

ATOM
Smallest part of a chemical element that retains its own character.

BILLION
One thousand million (10^9).

BLACK HOLE
A region of space-time that has a strong enough gravitational field to prevent even light from escaping. It appears that most galaxies have a supermassive black hole in their centers.

CCD CAMERA
Stands for Charged Coupled Device. The CCD captures light and converts it to digital data that is recorded by the camera.

CNO CYCLE
Standing for Carbon-Nitrogen-Oxygen, the CNO cycle is one of two nuclear fusion cycles that convert hydrogen to helium in stars, the other being the PROTON-PROTON CHAIN REACTION.

CONSTELLATION
A group of stars that appear to be near each other in the sky and form a recognizable pattern. In reality the stars may be at radically different distances from us and not connected to the rest of the group. Since 1930 the International Astronomical Union (IAU) has recognized 88 constellations. A few of them are known outside astronomical circles, for example: Orion, Cassiopeia, and the Big Dipper (or Plough), part of Ursa Major.

DARK MATTER
An invisible type of matter believed to exist as a result of observing known gravitational effects on bright matter.

DARK NEBULA
An interstellar dust cloud that obscures light from objects lying beyond it.

ELECTROMAGNETIC (EM) SPECTRUM
The light our eyes can see is only a small part of the overall EM spectrum which extends from gamma rays at the shortest wavelengths to radio waves at the longest wavelengths.

ELECTRON
A stable ELEMENTARY PARTICLE that is a constituent of all atoms, visualized as moving around the central NUCLEUS in a series of shells. The electron has a mass of 9.1×10^{-31} kilograms.

ELEMENT
A substance that cannot be split up into simpler substances. 118 elements are currently known, 94 of which occur naturally and 24 have been synthesized in the laboratory.

EMISSION NEBULA
A cloud of ionized gas glowing with high-energy photons, usually a site of formation of new stars.

GAMMA RAY BURST
The most powerful explosions in the universe: their origins are not known, but they may be extreme SUPERNOVA outbursts (hypernovae), the core collapse of massive stars to form BLACK HOLES, or the merger of NEUTRON STARS to form black holes.

GLOBULAR CLUSTER
A spherical group of tightly packed old stars that orbit the galactic center, the point about which the Milky Way Galaxy is rotating.

ELECTROMAGNETIC SPECTRUM

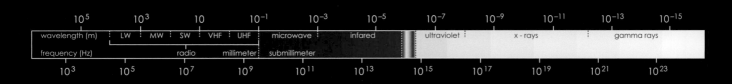

HI/HII REGIONS

Clouds of hydrogen in the galaxy. HI regions contain neutral hydrogen and the clouds cannot be seen in visible light, but can be studied with radio telescopes since they emit radiation at radio frequencies. HII regions contain ionized hydrogen, usually in the presence of hot stars, and it is the process of recombination of ions and free electrons to form neutral hydrogen that emits the light which we can see.

HERTZSPRUNG-RUSSELL DIAGRAM

A scatter plot of stars showing their luminosities versus their spectral classifications, first created by Ejnar Hertzsprung and Henry Norris Russell in 1910.

HOT JUPITERS

A class of gas giant exoplanets that are inferred to be physically similar to Jupiter but that have very short orbital periods. The close proximity to their stars and high surface-atmosphere temperatures led to their name.

INTERSTELLAR MEDIUM

Matter that exists between stars in a galaxy.

INVERSE SQUARE LAW

A law of physics which states that a specific physical quantity is proportional to one divided by the square of the distance from the source of that physical quantity.

IONIZATION

Any process that converts neutrally charged atoms or molecules into IONS, which have a positive or negative charge, through the addition or removal of one or more electrons.

KELVIN SCALE

A scale of temperature where the unit 1 kelvin (1K) is equal to 1 degree Celsius (1 °C), but the kelvin scale starts at -273.16 °C, absolute zero, the coldest temperature theoretically possible.

KUIPER BELT

Named after Gerard P. Kuiper. A disk of small solar system objects lying beyond the outermost planet, Neptune, and stretching out to about 50 Astronomical Units from the Sun.

LIGHT-YEAR

The distance traveled by light in one year, at a rate of 3×10^8 meters per second. It is equivalent to 9.5 trillion (thousand billion) kilometers or 6 trillion miles.

MAGELLANIC CLOUDS

The Large and Small Magellanic Clouds are galaxies in the Local Group of galaxies with our own Milky Way Galaxy. They are prominent in the night sky at southern latitudes.

MAGNITUDE

Brightness of an astronomical object. APPARENT MAGNITUDE 1 is 100 times brighter than magnitude 6, each magnitude being 2.512 times brighter than the next. Magnitudes brighter than 0 are minus figures; for example the brightest star in the sky Sirius is – 1.4 and our Sun is -26.8. ABSOLUTE MAGNITUDE is the visual magnitude that a star would have at a standard distance of 10 PARSECS.

MESSIER CATALOG

A catalog of 110 galaxies, nebulae and clusters compiled by Charles Messier (1730-1817). He was looking for comets, and noticed other fuzzy objects that might be confused for comets, and cataloged them M1, M2, etc. The designation remains in use today.

MILKY WAY

The luminous band of stars that stretches across the night sky. Because we are located in one of the spiral arms of our galaxy, when we look in towards the center of our galaxy we see a much greater density of stars than when we look out.

NEUTRON

A fundamental particle with neutral charge.

NEUTRON STAR

A stellar remnant that results from the collapse of a massive star following its supernova explosion.

NEW GENERAL CATALOG (NGC)

A catalog of nebulae and star clusters compiled by John Louis Emil Dreyer in 1888. Initially comprising 7,840 objects, two supplementary indexes were subsequently published adding more than 5,000 additional NGC objects.

NUCLEAR FUSION

The process in which two or more atomic nuclei fuse together under conditions of immense pressure and temperature to form a new nucleus and release energy. The process at the heart of the Sun is the fusion of hydrogen nuclei to form helium.

OPEN STAR CLUSTER

A group of a few hundred to a few thousand young stars that have formed from a molecular cloud such as an HII region and are weakly gravitationally bound together.

PARALLAX

The trigonometric parallax of a star is the angle subtended by one ASTRONOMICAL UNIT (AU) at the star's distance from the Sun, and is a measure of the star's distance. Observations are made at six month intervals at opposite sides of Earth's orbit around the Sun, so the baseline, the distance between the two observations, is twice the distance between Earth and the Sun, or 2 AU.

PARSEC

An astronomical unit of distance equivalent to 3.26 light-years. It is defined as the distance at which a star would have a PARALLAX of one second of arc (1/3,600th of a degree).

PHOTON

A particle of light-energy that can be transmitted at any wavelength.

PLANETARY NEBULA

An EMISSION NEBULA produced by the death of a star of approximately the same mass as our Sun. They appear as a glowing cloud of gas, usually in spherical or bipolar forms.

PROTON

A fundamental particle with a positive electric charge. The nucleus of a hydrogen atom consists of a single proton.

PROTON-PROTON CHAIN RECTION

Also called the P-P chain reaction, this is the nuclear fusion reaction at the core of our Sun that converts hydrogen to helium.

PULSAR

A rotating neutron star, often a strong radio source.

REFLECTION NEBULA

A cloud of interstellar dust comprised of fine particles that reflect light. We see them in the blue end of the visible spectrum.

SUPERNOVA

An explosion that can outshine an entire galaxy, caused by the reignition of NUCLEAR FUSION in a highly compact star such as a white dwarf, or the gravitational collapse of the core of a massive star.

SUPERNOVA REMNANT

The glowing cloud of interstellar debris from the explosion of a massive star.

TRILLION

1,000 billion (10^{12}).

WOLF-RAYET STAR

Exceptionally hot (of the order of 100,000 degrees at their surface) blue-white stars with spectra containing bright emission lines as well as dark absorption lines. They appear to be surrounded by rapidly expanding envelopes of gas.

ZODIACAL LIGHT

A cone of light rising from the horizon and stretching along the ecliptic, visible only when the Sun is below the horizon. It is due to thinly spread interplanetary material near the main plane of the ecliptic reflecting sunlight.

Brian observing in the Lowell Observatory, Flagstaff, Arizona

This is the 24-inch refracting telescope that Vesto Slipher used in 1912 to measure the speeds at which other galaxies were traveling away from us. Years later his research formed the basis for the theory of the expanding Universe. In 1914, he also discovered that the Andromeda Nebula (as it was then called – now it is correctly known as the Andromeda Galaxy) was rotating and on a collision course with our own galaxy. *Credit: David J. Eicher*

ACKNOWLEDGEMENTS

I'd like to thank my family, Lynda, Chris, and Amanda Eicher, for their support and encouragement during the book writing process. And I am indebted to my friend Glenn Smith, who is always on hand with inspiration and advice.

I also want to thank the many exceptional photographers who contributed to this unique book. First and foremost, the amazing J.-P. Metsävainio and his unique images will amaze you in this book. No one has done what J.-P. has accomplished — careful, analytical simulations of stereo views of distant nebulae, based on an intimate knowledge of their physical composition. Needless to say, this book could not have existed without the incredible talents of our good friend. Thanks are also due to my pals at Lowell Observatory, including Jeff Hall, Kevin Schindler, Dave Schleicher, Brian Skiff, and Michael West, for providing some imagery associated with that glorious institution.

Moreover, many skilled astroimagers contributed their mono images to this book. I've known nearly all of them for years as friends and contributors to *Astronomy* magazine, and they represent the cream of the crop of deep-sky photographers.

These include the spectacular work of Rogelio Bernal Andreo, the Spanish-American astroimager whose wide-field, very deep images are simply incredible. Chilean astronomer and astroimager Yuri Beletsky produces the most amazing wide-field images of the southern sky, including mind-blowing pictures of the telescope domes of the European Southern Observatory. I cannot give sufficient credit to Adam Block and his collaborators, who for years have produced great images in association with the University of Arizona. If there's a significant deep-sky object, Adam and his friends have taken a great photo of it.

Boston-based physician Steve Cannistra has been an astro hobbyist for many years and also produces breathtaking, deep exposures of nebulae. California-based Ken Crawford, a driving force in the highly-prized Advanced Imaging Conference, has amassed a spectacular portfolio of color astroimages, some of which you will see here. A specialist in shooting interesting, obscure targets, including some incredible dark nebulae, Thomas V. Davis is renowned for his creativity in astroimaging.

Chicago-born Tom Diana has conquered the night sky and produced a great portfolio of nebula shots. Neil Fleming demonstrates that persistence pays off; his extraordinary shots mostly originate from the last decade. Another driving force in the Advanced Imaging Conference, R. Jay Gabany has produced some of the most breathtaking nebula shots ever seen.

The master of imaging filters, Don Goldman, founded the business Astrodon and ran it for many years before his recent retirement, and you'll be stunned at the quality of his imagery. The Austrian physician Dietmar Hager has also mastered the skies in his free time with a great portfolio of shots.

Few have the skill and mastery of color deep-sky imaging to match Tony Hallas, the California imager who for many years operated a color photo processing lab in Los Angeles. Jaw-dropping wide-field nebular imaging has originated from the great Australian photographer Terry Hancock, who produces mosaics unlike anything created by others.

My pal and fellow Wisconsinite Mark Hanson has captured multiple nebulae from a favorite dark-sky site in New Mexico for a number of years. California physician Daniel B. Phillips also maintains a very high standard of deep-sky imaging. The Austrian great Gerald Rhemann, a camera and telescope specialist, produces stunning images from the Alps and also from Namibia.

A relatively recent convert to astroimaging, New Jersey physician Derek Santiago excels in wide-field imaging and shooting nebulae. Joel Short, a pastor in rural Indiana, has produced some of the best recent shots of large nebulae and wide-field vistas of star-crowded scenes. And not enough can be said of the amazing Alistair Symon, based in Tucson, whose wide-field mosaics are among the best ever produced for a large number of interesting fields filled with stars and nebulae.

I feel very privileged to call myself a friend of each of these imagers, and to include their amazing work in this volume.

On behalf of my co-author Dr Brian May and publisher Robin Rees, we would like to thank our friend Professor Derek Ward-Thompson for reading the proofs of *Cosmic Clouds 3-D*. And finally, we offer our eternal thanks to the sadly-missed Sir Patrick Moore, without whose inspiration this book would certainly never have happened.

D.J.E.

Stereoscopic Viewing

If you're familiar with the technique of **parallel free-viewing** of 'side-by-side' stereo pairs, you will have no trouble seeing all the 3-D images in this book. But to get the best immersive effect, use the OWL viewer supplied here.

This viewer is the new hand-held version of the acclaimed London Stereoscopic Company OWL Stereoscope. It's designed to give you high quality 3-D viewing of the stereo pairs in these pages.

Directions

Remove your LITE OWL viewer from the pocket opposite, and hold it with
either hand, locating the thumb on one of the grips near the lower corners.
Position the book in good light, with no shadows on the page.
Bring the lenses of the viewer close to your eyes, and position your
head squarely in front of the image pair you've chosen to view.
Move to a distance of about 5 inches (120 mm) from the page, and relax the eyes.
Don't squint or strain – you won't need to – and forget the idea
that you're viewing something close-up. Instead, expect to witness
a view of a distant scene through a window, or through binoculars.

Gently move back and forth a few millimeters until the image is in focus,
and slightly adjust the tilt of your head if necessary to align it with the page.
At this point the eyes should settle into a relaxed position as the full beauty of the
three-dimensional image appears. If you still see two flat images, don't give up!
Just disengage for a few seconds and look around the room, to remind yourself what
that feels like, and then try again, remembering to relax and look into the distance.
It will be worth it!

For more help, and information on all kinds of stereoscopic matters ancient
and modern, please visit our London Stereoscopic Company website at

www.LondonStereo.com

THE

London Stereoscopic Company, LTD.